OZ CLARKE
ENGLISH WINE

OZ CLARKE
ENGLISH WINE

FROM STILL TO SPARKLING

THE *NEWEST* NEW WORLD
WINE COUNTRY

PAVILION

READERS' NOTE

• The obvious title of the book should be 'British Wine', to cover wines made throughout the kingdom, but this term still refers to wines made from imported juice concentrate. They are usually fairly sweet and of pretty basic quality, so make sure you look for the terms English or Welsh wine on the label. The title of this book is 'English Wine' – although I have included a small section on Welsh wines. To say 'English and Welsh' wine several times a page through the book seemed to me to be just too cumbersome, so I have stuck to the simpler 'English' wine. I hope my Welsh friends will understand.

• The vineyards and wineries profiled from page 64 onwards cover personal experiences and reflections from my years of visiting and tasting up and down the country. This is not a comprehensive handbook full of statistics nor a who's who of top producers nor even a list of my favourites – I have tried to be representative and include the obvious 'big' producers as well as some 'more boutique' newcomers and my apologies to the many excellent producers I did not have space for in this book.

• Vineyard and winery profiles: Each entry includes a few essential details, to encourage you either to visit or to buy the wines and try for yourself. There has not been room to list all the many top quality tasting and food opportunities now on offer at many of these producers and the extraordinary variety of events and festivals available year round – I hope you will go and see for yourself.

• Symbols: sparkling and still wine produced.

• *Oz recommends* These are suggested wines to try and my favourites at the time of writing. They may not be their supposedly best wines but they are wines I have enjoyed and found interesting.

✳ refers to sparkling wine.

First published in the United Kingdom in 2020 by
Pavilion
43 Great Ormond Street
London WC1N 3HZ

ISBN 9781911624158

A CIP catalogue record for this book is available from the British Library.

10 9 8 7 6 5 4 3 2 1

Illustrations by Jay Cover

Reproduction by Rival Colour Limited, UK
Printed and bound by Toppan Leefung Printing Ltd, China

www.pavilionbooks.com

Previous pages: The South Downs in the distance, the morning mist in the valley, and Nyetimber's vines waking, ready for another day.

CONTENTS

HELLO FROM OZ

If I were writing this in October 2018 I would be gazing out of my window at bright golden sunshine still bathing the trees in warmth as a gorgeous summer resolutely refused to let go its sweet grip and fade into winter. I would be chatting to happy winemakers whose vats of juice from superripe grapes were overflowing and whose wineries were heady with the scent of wonderful flavours being created. Ah, the joys of climate change. It will be like this every year from now on, won't it?

But I'm writing this introduction in a very different October. The skies have been heavy and glum for well over a month, so that there's really no point in holding on to see if a balmy Indian summer will somehow bring back the cheer. There is a little sun peeping through – pale, nervous, barely enough to shift the autumn dew off the grapes as the bands of pickers don gumboots and waterproof jackets and trudge out into the muddy rows to judiciously harvest the healthiest and juiciest bunches of this year's considerable crop. Ah, the joys of being a marginal wine area, on an island buffeted by weather systems on all sides. No one has ever truthfully said that British weather can be predictable (no, not even the climate pessimists). And if we think that climate change is just about blue skies and record sunshine – well, that is part of the story, but only part.

Even if the 2019 vintage was one of the wettest for some time – we have had a tremendous run of warm, dry Octobers this century to make up any summer shortfall – well, the spring and the summer of 2019 were full of good things. There were no spring frosts to worry about. The vines flowered in good, warm, dry conditions and there were heatwaves that broke records in June and August. I spent quite a bit of the summer travelling through the English and Welsh vineyards and I didn't have one day when the sun wasn't shining, not one day when I didn't think what a glorious country we live in, and what an uplifting addition to our landscape is made by all our vineyards.

But there were also record cold temperatures in June and August, often right next to the heatwaves. And as the hurricane season in the Atlantic built up to being one of the most furious and vicious on record, with Hurricanes Dorian, Humbert and Lorenzo causing destruction on the other side of the Atlantic, the storms' tails hit the British Isles with unusual force, blocking out the sun and drenching our fields.

But it shows how far we have come as a wine-producing nation that we were ready for the tribulations of the year and, given the heat of much of the summer, large crops of good grapes could still be brought in, healthier, at higher sugar levels and weeks earlier than would have been possible a generation ago. We now have such a confident bunch of grape growers and wine producers in this country that problems are met head on, and quality is merely different, rather than worse, in most years.

Does this show we've come of age as a wine producer? I'm sure it does. The UK is still marginal in climate terms – but that's what you need to be if you're going to make great sparkling wine and fresh, aromatic still wines. Cool climate, yes, but warm enough to ripen Chardonnay and Pinot Noir for sparklers, and not too warm to bake away the hedgerow scent of the Germanic and hybrid varieties that still produce some of our most delightful still wines. The greatest marginal cool climate wine country in the world? Could we be that? Yes, we could. The men and women growing the grapes and making the wine here are as talented and passionate as those anywhere across the globe. And the raw material they have to deal with, the grape harvest they bring in each year, is intriguing, unique, sometimes unpredictable but bursting with the potential to make wines unlike any others.

And we must play our part. We must choose to drink English and Welsh wines. If we live in wine-making counties, we must support our local producers. And we must visit them. Wine tourism is of ever greater importance to wineries and vineyards. As more and more people seek 'experiences', not just the mere flavour of a bottle of wine – well, there's no better way to get the best out of a bottle than by visiting the place the grapes are grown and meeting the people who do the work. That's how you get to love and understand wine. Other countries have made themselves experts at this 'Wine Experience'. Now it's our turn, and there's no more beautiful country in the world to do it in than our own.

EXCITING TIMES

England probably has a 2000-year history of making
wine, but the trouble is that it's an inglorious one
– until now. And for the first signs that 'things can
only get better', we wouldn't have to go back much
more than 20 years to find the first flash of brilliance
which would transform a woebegone, unconfident
and, frankly, unnecessary English wine world into
the thrilling place that it is now – so full of potential
that I sometimes call England 'The Newest New
World Wine Nation'.

HOW IT USED TO BE

This Brave New World of English wine is based on bubbles, but the preceding 2000 years were not. Oh, there probably were some wines with a bit of prickle to them, because when the autumn was cold wines couldn't always finish off their fermentation and when you drank the wine in the following spring, it might have had a bit of a sparkle for a week or two as it finished its fermentation. But that would be chance and in any case was there really much wine being made? Did the Romans plant vines? A few archaeological digs imply they did, but it seems more likely that they imported boatloads of wine from the warmer areas of Europe that were part of their empire rather than struggled to make something decent from the shivering straggly vines they'd manage to grow in Britannia.

And after the Romans came the Dark Ages when the English gave up trying to be civilised for a few centuries and anyway, temperatures appeared to drop so there was even less incentive to try to ripen grapes. I'd sort of expect that drinks like mead would have been popular. This went on until about 1000AD, when things perked up a bit. We entered what the climate experts call 'the Medieval Warm Phase', which wasn't any warmer than in Roman times, but we had a Christian Catholic Church by then, and they needed wine for Mass, presumably red wine, the grapes for which they still would have struggled to grow, in chilly England. And we had a fair number of nobles keen on the finer things of life, many of them descended from the Normans who came over to England with William the Conqueror in 1066. The Domesday Book of 1087 showed that England – mostly London and East Anglia – had just 42 vineyards, few of them larger than an acre or two. Not a lot? Exactly.

You can try as hard as you like but it's difficult to find much evidence for a flourishing vineyard scene. Anyway, the south-west of France, including Bordeaux, became part of the English Crown in 1152 when Henry II married Eleanor of Aquitaine. A nice Bordeaux rouge? Or a thin sharp Essex white? You wouldn't choose the Essex white. Even literary references seem to favour chat about red wine – and that wasn't locally grown.

AND THEN IT GOT COLD AGAIN

The 'Little Ice Age' wouldn't have seemed so little at the time. It lasted from about 1400 to about 1850. The River Thames would freeze over, sometimes for up to two months, and frost fairs and markets were held on the ice. There were a few vineyards planted in places like Deepdene and Painshill in Surrey, but they mostly seemed to be rich men's follies, and, despite the effects of the Industrial Revolution beginning to warm things up a bit after 1850, honestly, not much of interest had happened by the time 1950 came along a century later. Between the First and Second World Wars not a bottle of English or Welsh wine was commercially produced. And it still wasn't warm.

A visionary called Barrington Brock established a viticultural research station at Oxted in Surrey after the Second World War. His vineyards were high up and the weather was usually cold and wet. But over the next 25 years he did at least prove that you could grow vines outdoors in the UK and make wine from them, even though he never managed to turn a profit for himself. Two other pioneers in the 1950s and '60s – Major-General Sir Guy Salisbury-Jones at Hambledon in Hampshire (now gloriously revived) and Jack Ward at Merrydown in Sussex – also cautiously planted vines and made wine. Not much. Indeed, hardly any. In 1964 the total national crop was recorded as 1500 bottles! (The UK produced 13.2 million bottles in 2018.) But they offered the wine for sale and someone bought it. The first faltering steps toward the modern glittering English wine scene had been taken.

STILL WINES, NOT SPARKLING

Both Guy Salisbury-Jones and Jack Ward believed in the eventual success of sparkling wine here, but this first revival was on the back of mostly thin, mean whites, often drunk without complaint for fear of offending your host. I had a few and, gosh, they were hard work. I kept hold of a 1976 from Chilsdown near Chichester for 35 years – regularly holding it up to the light to see if its bleach white colour had matured at all. It never had. So in 2011 I cracked it open, and this pale, unfriendly liquid forced its way out, indignant at having been disturbed and as lemon-lipped and cantankerous as it had been at its inception. Proud, unbending – yes. Fun to drink – definitely not.

The wines certainly became more drinkable during the 1980s, because slightly sweet, fruity German wines were all the rage in Britain – at one time half the wine we drank in Britain was German. Several German or German-trained winemakers turned up in England, at such wineries as Lamberhurst, High Weald and Tenterden; they took one look at our vineyards – most of which were full of German, cool climate vine-crossings like Müller-Thurgau – and thought, we know what to do here, make German-style wines. Germans made their wines sweeter by adding some grape juice – full of sugar – which they called Süssreserve. In Germany this usually made for a fairly flat mouthful, but England was cooler than Germany, the acid in our grapes was higher, and so a splash or two of sweetening grape juice merely balanced the acid rather than flattened the wine.

There were a lot of pretty poor wines made in the 1980s, but also some good ones, and you could reasonably say that making English Liebfraumilch-type wines was the first real sign of a new wave. I look at my tasting notes from the 1980s and early '90s, and find lots of charming, fresh, slightly leafy, slightly grapy white wines coming from wineries such as Staple, Barnsole, Tenterden, Biddenden or Syndale Valley in Kent, Lamberhurst, Carr Taylor, Nutbourne or Breaky Bottom in Sussex, Wootton, Pilton Manor or Three Choirs in the south-west and Pulham St Mary in East Anglia. All nice but none of them world-shattering. None of them doing something unique and better than anybody else. That didn't happen until the 1990s. You can call this the Second New Wave. You can call it a birth, a rebirth, an apocalypse. I call it 'The Nyetimber Effect'.

THE NYETIMBER EFFECT

In 1988 a couple of marvellously cussed Americans from Chicago decided that you could make great sparkling wine in an area of England with very similar soils and climate to Champagne, using the same grape varieties as Champagne, and the same methods and top-end equipment as were used in Champagne. All the experts said you couldn't do it – plant apples like everyone else. But every time someone told Stuart and Sandy Moss they couldn't turn their gorgeous medieval Sussex estate of Nyetimber into a world-class sparkling wine producer, they became more determined. As Stuart said, they persevered almost because it was so damned hard.

They made their first sparkling wine, out of Chardonnay, in 1992. It won the Trophy for Best English Wine in 1997. They made their second wine, from Chardonnay, Pinot Noir and Pinot Meunier, in 1993. This won the Trophy for Best Sparkling Wine in the World in 1998. I remember the shock, the excitement, the shivering thrilling realisation that I was tasting something entirely new, of astonishing potential, which would change my wine life for ever. Such events don't come round often in an entire lifetime of wine. I was grateful to be there at the start, and I knew that England had found its vocation as a wine country.

I had already been making speeches about the effects of climate change since the early 1990s – to deaf ears, frankly. But now it all made sense. The Champagne region was only a couple of hours' drive south of Calais on the English Channel. I had seen reports showing how their soils and many of the soils in southern England were the same. I knew that Champagne traditionally was about 1°C warmer than southern England, and experts said that was the difference between northern France just being able to ripen the classic French grapes, and southern England being unable to do so. Yet Champagne had been warming up all through the 1980s and '90s. Champagne in the growing season at the end of the 1990s was 1°C warmer than it had been a generation before. And if so, was southern England still only 1°C cooler than Champagne? Didn't that mean that by the 1990s, Kent and Sussex were just warm enough to ripen Chardonnay and Pinot Noir – and if someone took the plunge, could these chalky and sandy soils of England's south-east be a new Champagne?

So thank you to the Mosses and thank you Nyetimber. There really was a 'Nyetimber Effect'. It really did transform English wine. Nyetimber aimed for the very top of one particular quality pyramid, the sparkling wines of Champagne, which had been unchallenged in the world for several hundreds of years. Now there was a challenger – just over the English Channel. Nyetimber weren't the first to make fizz in England and to believe in its future. The really early pioneers of the 1950s made a tiny bit. Carr Taylor made some from Sussex Reichensteiner in 1983 and it was quite good. New Hall actually made a little Pinot Noir fizz in 1985 – but these were merely ripples. Who would take the big plunge? Nyetimber, with its wonderfully 'can-do' American owners who simply would not be denied.

Of course, there continues to be a lot of still wine being made in the UK, but we are increasingly seen by the rest of the world as sparkling specialists. About 70 per cent of our wine is sparkling, and every year the percentage of grapes being grown for fizz increases. But Britain is lucky in several ways. Champagne hardly makes any still wine – it lives or dies by its expensive sparkling wine and its hard-won reputation. As Britain sets out to share this sparkling glory, with all the investment and long-term planning that requires, still wines – cheaper to make, cheaper to sell – provide a crucial cash flow safety valve. While new plantings career onward and upward, as they have done since 2017, still wines actually offer one of the few channels, along with wine tourism and hospitality, to get some money coming back into the business.

UK PRODUCTION FIGURES 2012–2018

(shown in millions of bottles)

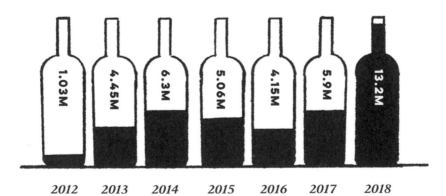

| 2012 | 2013 | 2014 | 2015 | 2016 | 2017 | 2018 |

1.03M 4.45M 6.3M 5.06M 4.15M 5.9M 13.2M

This shows that production is still erratic in the UK, but the direction will be relentlessly upward as new vines come on stream.

But the spirit is a sparkling spirit. Ridgeview quickly joined Nyetimber in the 1990s as a winery solely concerned with making top Champagne-style fizz. Chapel Down, the winery now occupying the old Tenterden site, is a major fizz producer, and almost all the new tyros are focused on fizz – Gusbourne, Coates & Seely, Hattingley Valley, Hambledon, Exton Park, Greyfriars, Rathfinny, Harrow & Hope, Furleigh and Simpsons – many of them make some still wine but all of them concentrate on fizz. And did I mention Taittinger and Vranken-Pommery? They are two leading Champagne producers. They have both established estates – in Kent and Hampshire respectively – with the objective of making some of the best sparkling wine in the world. They won't be the last ones to cross the Channel.

CLIMATE: IS IT ALMOST PERFECT NOW?

Honestly, if you had to choose a nation in which to grow vines solely based on its geology and soils, Britain – England, Wales and even Scotland – would be hard to beat. So why haven't we had a thriving wine culture for the last few hundred years? Well, for those of us who live here – the weather. The Roman writer Tacitus dismissed Britain as a filthy, foggy, rainy hellhole 2000 years ago. What? England? The south and east of England? In the 21st century? Filthy, foggy and rainy? I don't think so. In fact, millennials, and those even younger, could be excused for not recognising this view of the British Isles at all. People who grew up in the 1950s, '60s, '70s and '80s – ah, yes, they remember the fogs, the drizzle, the certainty that August Bank Holidays would be miserable. Do older people always look back and say things were different then? Yes, they were. They were worse.

Climate change and global warming are probably the biggest challenges the human race faces today. In most parts of the world, the effects of climate are already worrying and will probably become catastrophic. There are just a few corners of the globe where climate change is having a positive effect. And if I had to choose one place where climate change has completely transformed a way of life for the better, it would be in the vineyards of England and Wales. That transformation isn't without its challenges.

But let's have a closer look at this new British climate which promises so much.

IT'S WARMING UP

First, and most important, it is warmer. It's warmer in Cornwall. It's warmer in Sussex and Kent. It's warmer in Essex and Norfolk, and Monmouth and Conwy and even Yorkshire, for goodness sake. Which means that grapes will ripen in areas they never would before, and that varieties like Chardonnay and Pinot Noir, which need more warmth than England has had for 2000 years, are now the dominant varieties in our southern vineyards and performing superbly. But how is it warmer? Well, there's no question that we have been getting considerably more very hot days in our summers. Very hot for Britain I would describe as 30°C and more. In the last couple of decades of the 20th century, many years would pass with not a single day reaching 30°C. In the first 20 years of the 21st century, only 2007 saw a summer with no days at 30°C, although the average temperature for the whole of 2007 was the second highest on record.

And this heat is extending back into spring and forward into autumn. Record heat is being recorded in June, May and April, let alone March and even February (I have more than once recently sat outside my local pub with a pint in February – in shirtsleeves). August's heat is stretching into September (in 2019 they were still recording unheard-of temperatures as high as 33°C in south-east England at the end of August: on the Bank Holiday for a start). September now rates as late summer, and frankly October often does as well. Three of the five warmest ever Septembers have been in this century, and nine of the ten hottest Octobers have been in the last 20 years. Put this together with much more professional vineyard management and you have vineyard owners boasting that they are picking grapes with twice as much sugar as a generation ago, and the harvests are up to a month earlier. In 2018 the English harvest started on August 28.

But this astonishing change in our summer weather is not an isolated phenomenon. We are surrounded by various weather systems in Britain as a maritime nation. They are becoming more extreme, too. In particular, as the systems to the west warm up, and the south-westerly is the prevailing wind for the majority of our vineyards, we get more rain events, much stronger winds and greater likelihood of things like flash flooding, torrential downpours and hail. Often we see a localised area getting several months' worth of rain falling in a single day.

Imagine if your vineyard is in the middle of one such area. Periods of great heat can often bring a much more violent reaction of storms, and even cold (look how the weather went from very hot to very cold at least twice in the summer of 2019, during June and during August – after which, of course, record heat was once more recorded). This climate change phenomenon is a roller coaster ride. Few experts any longer simply refer to it as 'global warming' and many actually call it 'climate chaos'. And they're not wrong. Nothing is predictable any more. At the moment Britain's vineyards are on an upward curve. Let's make the most of it while we can.

And there's one more thing. Frost, especially springtime frost. Again, this isn't just a British phenomenon. Europe is increasingly being hit by vicious late spring frosts. Famous vineyard areas like Burgundy in France have been hammered again and again in the 2010s. If you have a cold winter and the vines are late to wake and start their growing season, with rising sap and buds being pushed out, a spring frost may well do no harm – the vine's branches can cope with all but the worst frosts at way below 0°C. But if you have a warm February and March, as we now often do, then by mid-April those buds may well be pushing out and opening up, encouraged by the sun. So a frost that might have been harmlessly rebuffed by tough vine wood can devastate the delicate vine buds and decimate your crop.

On April 26 and 27 2017, a terrifying frost destroyed crops all over Europe. Southern England's lovely warm March and early April weather had encouraged the vines to push out their buds. Overnight temperatures as low as –7°C meant many vineyards lost up to 80 per cent of their crop. So much for global warming.

LOCATION IS KEY

They say that the most important factor when it comes to selling a house is location, location, location. In a marginal vineyard country like Britain, the coolest of any of the serious wine-producing nations, suitability of your vineyard site must be at the top of your list of priorities. Well, you'd think so. But an amazing number of British vineyards in the second half of the 20th century were planted simply because the piece of land concerned was available. Often it was part of a property owned by a family who thought the idea of a vineyard was a lovely one, especially after a series of delightful summer holidays in places like France.

Often it was a patch of farmland that wasn't being used for much, so why not try vines? And often the plot chosen was on heavy, wet, cold clay soils of the sort that England is full of, and if there is one thing vines don't like it's having cold wet feet. They don't much like sitting there in the teeth of a gale either, and many early sites were completely exposed to the elements. And vines do like a fair bit of sun, so those early sites which were north facing were hardly likely to give you a decent drop of remotely tasty wine.

WHAT IS A GOOD VINEYARD SOIL?

England is brilliantly furnished with good to great vineyard soils which drain well and offer the vines the nutrients they like. One world vineyard authority, Dr Richard Smart, says that the Paris Basin is the motherlode of all vineyard land. So what is the Paris Basin? It's the ring of chalk and limestone ridges that run around Paris. It passes through Champagne and then it moves north from Calais across the Channel to the cliffs of Dover and England. In southern England it includes the North and South Downs. The chalk also extends right up through the Chiltern Hills to Hertfordshire, Cambridge, Norfolk, Lincolnshire and East Yorkshire. Jurassic limestones spread from Dorset up through the Cotswolds and on into Lincolnshire and northern Yorkshire. This pale soil is very well drained and reflects heat back onto the grapes. Sounds perfect.

Facing page above: the splendid chalky isolation of Rathfinny's vines and flint cottages in the South Downs.
Below: Hampshire is all about chalk and this view of cellar excavation at Exton Park shows there is almost no topsoil at all.

And this is just the chalk and limestone-based soils. The Thames Valley and elsewhere also have warm, well-drained gravel beds great for vines. The greensands and sandy clays that keep popping up in Kent and Sussex and elsewhere are warm and well-drained and highly suitable for vines. Another great climate expert, Dr Alistair Nesbitt, reckons there are well over 30,000 hectares (74,100 acres) of prime vineyard land available in England – and barely any of that has been planted yet, even by estates which are situated on the chalks, the limestones and the greensands.

And there are all the other soils. The most common of these in southern England is clay, for example Wealden clay or London clay. Clays mixed with limestone, gravel or sand can make excellent vineyard soils. Pure clay is much more difficult. Tough, thick, claggy clay is cold and waterlogged in winter and wet weather, and cracked and parched in drought. It doesn't sound promising, yet, especially in Kent and Essex, delicious wines are being made from grapes grown on the clays, even including what are often England's best Pinot Noirs and Bacchus. How? Well, this is where good old climate change and local weather go hand in hand. Essex is the driest part of England. Suffolk, Norfolk and Kent do pretty well, too. And there are an increasing number of places like the Crouch Valley in Essex which are proving to have amazingly suitable local climates. It's rarely going to be drought-dry there, so the clays are just warm enough, not too damp, and delightfully soft, scented red and white wines are possible. On heavy clays that shouldn't be possible. In Europe it normally isn't. Somehow it can work in England, and above all, in Essex.

And as the climate warms, even regions like Scotland and the Welsh borders are revealing soils that are pretty much the same as those in Germany's Mosel Valley, France's Beaujolais and northern Rhône Valley, and Portugal's Douro Valley. Dr Richard Selley, a greatly respected British geologist – and wine enthusiast – put together a scenario of vineyards in Britain based on widely accepted projections of temperature increases up to 2080. The Thames Valley and Hampshire could be too hot for wine grapes; our best Chardonnay and Pinot Noir could be coming from the uplands of Derbyshire and Yorkshire, and refreshing whites could be made on the banks of Loch Ness.

MAIN VINEYARD LOCATIONS IN ENGLAND AND WALES

Vineyards are cropping up in increasing numbers all over England and Wales, from Yorkshire down to Cornwall, but the biggest concentration is in the drier and warmer areas of East Anglia and the South-East.

SITE SELECTION

Britain's leading vineyard expert, Stephen Skelton, thinks that the soil is less important than the site being protected from the worst of the wind and rain, and being well exposed to the heat of the sun. There has been a rush during the 2010s to plant the chalky downland of southern England because the soils seem identical to many of those in Champagne, and global warming has meant that the climate is now fairly similar, especially in many parts of Kent, Sussex and Hampshire, to what it was like in Champagne at the end of the last century. So this change has been rapid and has had a significant effect on what types of site can now be successfully planted. The rule used to be that anything over 100 metres (330 feet) high wouldn't ripen grapes. It will now and there are vineyards at 200 metres (660 feet) high that are producing crops of good grapes. Many of these are on chalky sites.

Even so, they will generally get more wind than lower sites and that is a mixed blessing. Breezes are generally welcome in the summer and autumn because they keep the air moving and stop fungal infections from taking root. But strong winds lower the temperature, blow the flowers off the vines so they don't set a crop, and can even break the branches of young vines. You see a lot of windbreaks in British vineyards, usually in the south-west, to catch the worst of the gales. Sometimes even they are not enough.

Airflow can also help in preventing frost. Many sites have frost pockets in them where cold air drains and sits, freezing any vines that may be planted there. Keeping the air moving is the best way to combat frost, and various systems can be employed against frost, all of which rely upon trying to keep warmer air moving into colder spots. Chapel Down's Kit's Coty vineyard in Kent probably has the best system. The high speed Eurostar trains hurtle past the bottom of the vineyard several times an hour, at 200 kilometres (124 miles) an hour – wonderful for keeping the air constantly on the move.

And whether you're on a slope or not matters, too. In a cold climate like Britain's you want to catch as much of the sun as possible. Slopes, generally, do this best. Although in midsummer there isn't much difference in the amount of sun a flat or a sloping vineyard gets, as the season wears on and the sun drops lower in the sky, a vineyard facing south and planted on a slope of 30° will get far more direct sunlight. By October such a site may be getting 30 per cent more sunlight than a flat site. Most new vineyards are sited on slopes, although rarely as steep as 30°. Usually the objective is to have a southerly aspect toward the sun. Some growers prefer south-easterly so that the vines warm up more quickly in the morning. Some prefer south-westerly, since typically afternoon sun is warmer than morning sun. Surely no one would want to face north? Well, Peter Hall at Breaky Bottom in Sussex has a north-facing vineyard that always ripens before its southerly neighbour. Robb Merchant at White Castle in Wales has just planted a northerly slope, but he shows me the path of the sun, the wind protection from a copse of trees, the rain shadow from the local mountain, and he reckons he'll be able to ripen Cabernet Franc on it, something no one else has managed yet in Wales. And surely New Hall in Essex know everything about their local conditions after 50 years of growing some of England's best grapes? Their new vineyard below the tranquil Purleigh churchyard looks to have a fair bit of north-facing land to me.

PLANTING LIKE MAD: BRITAIN'S VINE EXPLOSION

Critical mass really does matter. And success breeds success. And a few really hot summers in Britain turn people's heads and fill their minds with dreams of owning a vineyard more than anything. 1976 was a spectacularly hot summer – I spent it acting on the melting tarmac streets of inner Manchester – I should know. Despite the summer coming to an abrupt halt at the end of August and two months of gloomy, drizzly English miserableness ensuing, there was a rush of interest in establishing vineyards across the UK, and some of the nation's most impressive vineyards were planted in the aftermath. They didn't all prosper, and despite the 1990s bringing some of the warmest ever British weather, including some decent autumns, there was a drop of more than a fifth in the vineyards' acreage leading up to 2003.

Ah yes, 2003. This was the year, far more than 1976, when people began to wake up to the fact that we are living in a warming world. Temperatures reached 35°C for the first time since records began nearly 400 hundred years ago. 35°C? Faversham in Kent reached 38.5°C on August 10. A big intake of breath, and then a renewed enthusiasm for vineyards, which saw them increase by about 20 per cent in 2006 alone. Since then warm winters, springs, summers and autumns have continued to fuel the belief that England's time has come and maybe Wales's, too.

And the proof is in the pudding. Beginning this modern era with the glorious wines of Nyetimber and Ridgeview at the turn of the century, a whole swathe of the 21st-century plantations are producing wines – generally sparkling – that seem to improve with every vintage. Nyetimber and Ridgeview are still there, stronger than ever, but there's an ever-growing jostling band of pretenders whose wines can equal theirs. Success breeds success. Critical mass matters. In 2017 a million vines were planted. In 2018 a million and a half vines went into the ground as the vines that were already there produced a harvest of 13.2 million bottles to eclipse the previous largest crop of 6.3 million bottles in 2014. Which spurred everyone on even further. At least 3 million vines were planted in 2019, maybe a lot more, as weather records tumbled once again, and several new 'hottest ever' days were recorded. Vineyard Number 9321 is one of the UK's newest registered vineyards – it's only two vines, in Upper Lodge's kitchen garden in Sussex – but from small acorns .... And where are most of the new plantings? On the soils and slopes that have proved themselves in the last decade. On the chalks of the North and South Downs, on the greensands and sandy clays of Sussex and Kent, in the protected calm of the East Anglia coast.

The heat of 2003 had an effect across the Channel, too. Champagne's reputation for the finest fizz in the world has relied on a regular supply of grapes creeping to ripeness in fairly marginal conditions. 2003 in Champagne produced their earliest harvest ever in blistering weather after a blistering summer. A lot of producers thought they would simply be unable to make proper sparkling wine – and if this was a sign of the future, they'd need to think about finding somewhere cooler to produce their grapes. You

INCREASE IN PLANTINGS 1975-2018

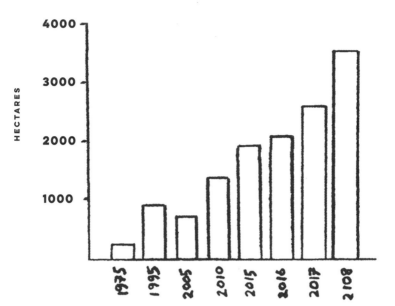

After a bumpy start, British vineyards are rapidly expanding, with most of the plantings being of the Champagne varieties: Chardonnay, Pinot Noir and Pinot Meunier.

didn't have to be a genius to realise that the most suitable soils and climate were just over the Channel in southern England.

THE FRENCH ARE COMING

I first became aware of the French interest when I started getting telephone calls in 2004 asking me to confirm this or that rumour that the big Champagne houses had moved in on Kent under hidden identities and bought large swathes of chalky, infertile, low value agricultural land. Low value if you're thinking of raising a crop of wheat. Remarkably high value if you're a Champagne producer looking for a safety net. The price of a hectare of Kentish chalky farmland was about 1 per cent of the price of good vineyard land in Champagne. These were only rumours, but I still look at swathes of Kentish land that I should have expected someone to start developing into vineyards ages ago – and it's just sitting there.

Have the Champagne houses already land-banked large chunks of England for the time when they are forced to look elsewhere for their supply of grapes? But it wasn't just rumours. In 2005 my dentist told me of a French friend of his, Didier Pierson, a grower from Avize, one of Champagne's best villages, who had planted a vineyard on a high, chalky ridge in Hampshire. In 2007 and 2008, as I was touring various vineyards in the south, I couldn't resist asking – have the Champagne boys tried to buy you yet? Every vineyard I went to had had a visit, some had had an offer but none had said yes – yet.

And then in 2015 the fox broke cover. Taittinger, one of Champagne's greatest houses, announced that they had purchased an old apple orchard on chalky soil near Canterbury in Kent, and they were planting the Champagne grape varieties – 20 hectares (50 acres) in 2017, another 8 hectares (20 acres) in 2019, with more to come. Their first crop was in 2018. I know where the wine is being stored, but I haven't had the audacity to ask to taste it. After all, Taittinger will want to make a big splash as they launch their first wine under the Domaine Evremond label (Charles de Saint-Evremond was an exiled French nobleman who performed wonders in encouraging Champagne drinking in England in the 17th century).

And they shouldn't tarry too long. Vranken-Pommery, another famous Champagne house, has bought an estate on the Hampshire chalky downlands near Exton Park. They had planted over 40 hectares (100 acres) by 2019 and have been burnishing their skills by already producing some sparkling wine called Louis Pommery England at a neighbouring winery, Hattingley Valley. The years of 2018 and 2019 were very hot in France. Will more Champagne houses bow to the inevitable and set up stall in the UK?

LET'S AVOID TOO MUCH REGULATION

The increasing interest of the French in English vineyards can only be a good thing for improving the reputation of the nation's wines, but it makes me wonder a little about whether an ever-increasing French and, maybe, other European presence in the UK, will bring about pressure to adopt European bureaucratic controls for British wines. Europe has a very tightly regulated and, frankly, restrictive appellation system for controlling places where you can grow vines, which varieties you can grow and what types of wine you can make. There is also a looser system of control for areas without much tradition or for grape varieties that are not usually grown there. In France we'd know these wines most frequently as Vin de Pays. So there's a straitjacket approach and there's a laissez-faire 'anything goes' approach. All the greatest traditional wines of Europe are made inside the appellation straitjacket. But the majority of the most exciting, imaginative New Wave, and often affordable, wines are made under the very broad umbrella of the laissez-faire.

Well, France has had 1000 years to figure out what brings the best results, so maybe it deserves to be able to lay down the law about what works best in certain regions and, I suppose, put a bit of a straitjacket around it. But Britain has no tradition of excellence at all. To be honest, everything is wonderfully experimental. We've got ideas as to what vineyard sites might work best. We've gone headlong into planting the Champagne varieties while keeping loads of other less trendy varieties in the soil. We *think* we know what's best for each of us, but we don't *know*. In this way, we are a new frontier, New World wine country with everything to prove, everything to explore. There are quality systems in place in the UK, but they are far less restrictive than those in France, and all the better for it. Any system which bureaucratically attempts to shackle our enthusiasm and our imagination can't be a good thing. We don't yet know what will excel, where, and how. But there is an impressive bunch of wine entrepreneurs determined to find out, and they don't need any help from the authorities.

Britain uses the European indicators of quality and authenticity in terms of regional original or traditional production – the PDO or Protected Designation of Origin and the looser term, PGI or Protected Geographical Indication. Many European wine producers, for example in France, Spain and Italy, use PGI when they are looking to make wines which don't conform to tradition. In Britain PGI is used widely because it has fewer restrictions.

GRAND CRU VINEYARDS
People have also started talking about 'Grand Cru' vineyards. Grand Cru means 'great growth' or 'great site'. The very best wines in Burgundy come from a very few great sites that have

earned their spurs over 1000 years. Bordeaux has a few such properties famous for hundreds of years. The Loire Valley, Alsace, Champagne, they've all worked out their special sites and accorded them 'Grand Cru' status. And now we are talking that language in Britain. People are calling Kit's Coty in Kent a 'Grand Cru', but it's only made a handful of vintages, impressive though they are. Gusbourne talk about developing a Grand Cru in their vineyards down by the Romney Marsh. Is the Tillington site near Petworth in Sussex a Grand Cru for Nyetimber? Will Denbies' south-facing, 40 per cent slope be Surrey's finest? We don't know. It's far too early to say. And with climate change accelerating, the probability is that the real 'Grand Cru' sites aren't even planted yet. There are a series of south-facing, protected horseshoe bowls in the South Downs – Wiston Estate has one and it might even have several. I look at them and say – wow, let's dream of 'Great Sites', let's seek them out and try to draw out the magic they may possess. But let's not make 'Grand Cru' a legal entity – if it's good enough, wine lovers will recognise it for what it is without any help from lawmakers.

A QUESTION OF STYLE: SPARKLING OR STILL?

Wine comes in dozens of different styles – from still white and red to varying degrees of pink, from orange, through sweet and fortified to sparkling. The tiny wine industry of England has come of age and produces world-class wines, especially fizz, which is now the most important category in England, but what about the still wines? They're still incredibly important and with the phenomenal expansion of vineyards may well become more important in the future – still wines are cheaper to make, quicker to bring to market and crucial for the winery's cash flow.

In the following pages I've outlined the styles, both sparkling and still, that are important for English and Welsh wine.

STILL AND SPARKLING WINES PRODUCED IN THE UK

(shown in millions of bottles, 2018)

69%

31%

9.1m bottles   *4.1m bottles*

Sparkling wine now easily outpaces still wine in Britain, as traditional-method fizz begins to earn a worldwide reputation for quality.

SPARKLING WINES

It is quite remarkable how Britain has gone from being a still white wine producer of declining relevancy in the world of wine to being a contender for the title of world's best sparkling wine producer – all in the space of a generation. Frequently nowadays we wine critics have to remind ourselves that still wine is important in Britain and that it has never been better, at the same time as more and more examples of fizz that are good to outstanding appear with every vintage, and awesome amounts of land get planted with vines each year – most of them with the three Champagne varieties, Chardonnay, Pinot Noir and Pinot Meunier, which now comprise 70 per cent of all the vines planted. And the objective is always to make sparkling wine, not still wine. No one has yet set out to make the world's greatest red Pinot Noir or white Chardonnay in Britain but quite a few people have set out to make the world's greatest sparkler.

This ambition is firmly rooted in making wines from the Champagne varieties in the same way as they produce them in Champagne. By inducing a second fermentation inside the sealed bottle, capturing the bubbles that fermentation produces. Most of the wines of Champagne are made from a blend of all three Champagne grapes, so that a typical 'Classic Cuvée' would probably have at least half of the base wine coming from the black Pinot grapes, pressed extremely carefully to avoid staining the juice red. The rest of the juice would be Chardonnay. This 'Classic Cuvée' is definitely the most popular style in Britain, too.

Initially, during the 1990s and 2000s, these wines would all be vintage-dated, from a single harvest. But the unpredictable nature of the British weather – particularly the uncertainty as to what kind of quantity of grapes the vines can produce – has meant that, just as in Champagne, more and more wineries are now making a non-vintage style using wines from more than one harvest – or, as Nyetimber, puts it, a 'multi-vintage' style. Several of the larger wineries such as Nyetimber, Hattingley Valley and Chapel Down, as well as smaller ones such as Exton Park, put a considerable percentage of their wine each harvest into a reserve. This reserve wine ages and softens and deepens in flavour and can then be added back into a blend of younger wines to achieve a balanced and consistent style which simply isn't possible with wines based on one vintage only.

VINTAGE, NON-VINTAGE AND LUXURY CUVÉES

Most English sparkling wine producers started out making vintage wine, with that year stated on the label, but most now produce both vintage and non-vintage. Vintage wine comes from a single harvest whereas non-vintage and multi-vintage wines are a blend of harvests. These require reserve wines to be put aside each vintage for later blending to soften and improve the flavours and to achieve consistency year on year. Since so many sparkling wine producers in England are recently established, reserve stocks have been low but each year should add to the stocks available and large vintages like 2018 and '19 will provide the perfect opportunity for wineries to build up their reserve stocks – so long as they have enough storage tanks. Luxury cuvées are special releases, using the best grapes and the best selection of base wines, often with barrel ageing and extended time in bottle on the lees. Some of these wines are superb but you pay for the privilege. If you want to see what they are all about, try the following: Chapel Down's Kit's Coty Coeur de Cuvée, Coates & Seely's La Perfide or Hattingley Valley's King's Cuvee.

BLANC DE BLANCS

The very first Nyetimber wine to achieve renown was a Blanc de Blancs, made only from Chardonnay grapes. This is one of the most popular – and highest-priced – styles of wine made in Champagne, and it looks as though this trend will be followed in Britain, too. The best Chardonnay grapes in Champagne come from one very chalky region called the Côte des Blancs, the White Slope. Nyetimber's 1992 actually came from sandy soils, and many of the best English Blanc de Blancs wines until about 2010 came from clay and sandy soils. Since then there has been a rush to plant southern England's plentiful chalk slopes with, above all, Chardonnay and more of these 'white grape only' sparklers do now come from chalky vineyards. There's no doubt that these have an arresting, focused, limpid quality and an acidity that sweeps across your tongue. The wines from grapes grown on clay and sand are generally rounder, softer, creamier, nuttier. Neither is better or worse, just gloriously different. And many top Blanc de Blancs wines will mix grapes from chalk with non-chalk grapes.

BLANC DE NOIRS

Blanc de Noirs is a less popular, but equally impressive style of wine made only from black (*noir*) grapes – in almost every case, Pinot Noir with or without some Pinot Meunier. These wines are fuller, broader, a little richer on the palate, the acidity isn't quite so taut, the texture isn't quite so lacy. But England is producing tremendous examples, even if producers find them more difficult to sell. That means they may be a bit cheaper to buy, remember. Quite a bit of the best Pinot fruit is grown on clay, some on greensands and sandy clays, but also some on chalk. We are at such an early stage in our development as a great sparkling wine producer that we're all still experimenting.

SPARKLING ROSÉ

If Blanc de Noirs whites are a difficult sell, frothy pink wines may be the easiest sell of all. Pink is right in vogue at the moment, and Britain does it very well. The wines hold onto their pinging English acidity, but wrap this with a fruit of pink strawberry, maybe cherry, maybe pink apple and cream. And these pink wines are not just made from black grapes. Many of the best have a significant amount of Chardonnay in them. Oh, and there are some sparkling reds, too. Not many, but a throatful of Ancre Hill's Triomphe Perlant – purple red, packed with rough black plum and sloe fruit rudely scented with balsam with a restraining collar of staining foam – should convince you.

HOW DRY IS DRY?

Traditionally Champagne has contained a certain amount of sugar or 'dosage' to balance the acidity and accentuate the fruit. Not much sugar, in fact you only really notice it when it's not there. There has been a trend for making bone-dry or Extra Brut Champagne – Champagnes labelled Brut nature, Pas dosé or Dosage zéro contain no sugar at all – and many of these wines make you realise how barely ripe the grapes were that made the wine. A little sugar helps. In England, the acidity in the grapes is generally higher than in France, but in a brilliantly piercing way. There are a few English wines with no sugar at all, and they are impressive though not always lovely. Just a few grams of sugar per litre – the magic figure seems to be somewhere between 6 and 9 grams (in Champagne these wines are labelled as Brut) – seems to me to create the perfect balance in English fizz and it's not really noticeable until you take it away. When the label says Dry or Brut you will get the proper sensation of a dry wine.

OTHER WAYS OF MAKING SPARKLING WINE

I've been talking about sparkling wines made in the expensive and time-consuming Champagne or 'traditional' method. This is how all the best examples are made. And in fact 99 per cent of English sparkling wine is made this way. However, there are some wines made by the cheaper 'tank' or Charmat method, a more industrial process which is banned in Champagne. Producers like Flint in Norfolk have shown that you can make tasty stuff in this way. The bubble is not quite so fine and long-lasting, but so long as you're not trying to create the so-called classic Champagne flavours of croissant, hazelnuts and crème fraîche, which rely on long ageing of the wine sitting in its bottle in contact with yeast lees, then the commercial reasons for it make sense. Grapes like Bacchus, Reichensteiner, maybe Solaris, maybe Rondo, which don't in any case pick up those creamy, nutty flavours, are quite well suited to the Charmat method. And you can release the wine younger and sell it for less than the traditional method wines. We haven't seen much Charmat method yet. Big vintages like 2018 will mean we'll see a lot more.

There's one even cheaper way to make wine froth a bit – you can simply carbonate it. Some of the worst sparkling wines in the world are made like this. Yet they can be a good drink, especially if the base wine is from a very well-flavoured grape variety. New Zealand makes some carbonated Sauvignon Blanc that tastes like – well, fizzy Sauvignon Blanc. Some people say Bacchus is Britain's answer to Kiwi Sauvignon. Maybe. Certainly, Chapel Down has started making carbonated Bacchus and it tastes like – well, a very easy-going, frothy Bacchus.

STILL WINES

DRY WHITE WINES

It's not often you find yourself saying thank goodness for Liebfraumilch. But it was the German style of fruity, slightly sweet white wines which gave British vineyards their first cautious wine boom. When the modern wine era began in the 1950s, any kind of template for style was likely to be French and consequently dry. That's OK if you have ripe grapes, but England in the 1950s and '60s, and frankly the '70s and '80s, didn't have properly ripe grapes. I've tasted a few of the dry whites made in the 1950s and '60s and, honestly, you couldn't finish a glass, let alone a bottle.

In the 1980s German wine, fruity, slightly sweet, became the most popular style of wine in Britain. The arrival of several German, or German-trained, winemakers in England who knew how to turn English grapes into a sort of Liebfraumilch meant that home-grown grapes, low in natural ripeness but dollied up with added sugar and the sweetly fruity unfermented grape juice known as 'Süssreserve', or 'sweet reserve', could make a reasonably attractive, vaguely grapey, sometimes scented, but rarely dry, white wine. But the 1990s were brutal to this rather neuralgic me-too Germanic style. Australian, Californian and New Zealand wines hit the UK market – ripe, alcoholic, sunny but dry. There was no way that England's home brew could match any of these, and wineries and vineyards began to go out of business.

And if it weren't for the carnival cavalcade of fizz that Nyetimber and Ridgeview lit the flame for in the 1990s, I can't see that we'd have that much to celebrate in English wine right now. Sparkling wine massively upped the ambitions of English winemakers. Helped by climate change, but also by far better vineyard management brought into play by the usually well-financed and sophisticated sparkling mob, the old varieties which used to make Germanic styles, but also the queen of whites – Chardonnay – could achieve a ripeness level that didn't need you to add any sugar to the wine. The new goal is to make great Chardonnay, but the legacy grapes like Seyval Blanc, Ortega and, above all, Bacchus can now ripen well enough to make wines that don't need extra sugar, and can, I suppose, be likened to New Zealand Sauvignon Blanc.

MEDIUM WHITE WINES

Medium and medium sweet wines are pretty out of fashion nowadays. But they are a very attractive style when your grapes are high in acidity. Too much acidity makes a dry wine raw and harsh. But if you leave some sweetness in the wine, it balances the flavour to such an extent that the acid can merely seem like a squirt of lime or lemon juice, freshening up the fruit, and the sweetness becomes barely noticeable. In cold countries it's often not possible to ripen grapes enough to lower their acidity to easy-going levels. Britain is getting warmer, but as viticulture stretches northward toward Yorkshire and westward into Wales many vineyards struggle to reduce the acid in their grapes. In which case, medium styles make absolute sense. Grape varieties like Solaris and Orion are often made with sweetness remaining, and are much better for it. Varieties like Huxelrebe and Siegerrebe aren't common but make attractive, sappy, zingy medium styles. And even Bacchus is often made with quite a few grams of sweetness left in the wine – but this balances the acid very attractively, and we'd never know we were drinking a medium wine.

SWEET WINES

There are two ways to make a naturally sweet wine – one that isn't fortified with spirit as with port. If your grapes are attacked by a particular fungus called 'noble rot', which intensifies the sugar without turning the juice into vinegar, you can make great sweet wine. Sauternes from the Bordeaux region of south-west France is the most famous example. Or you can leave the grapes to freeze on the vine and press off the thick, sugary goo while the water is still frozen as ice. Canada, not surprisingly, is famous for its icewine.

Neither of these styles of sweet wine is common in Britain. You need very ripe grapes in the first place for the noble rot fungus to strike your crop, rather than all the other unwanted types of rot. In the old days this occasionally happened by mistake, but apart from Denbies, who have a plot of Ortega that gets hit by noble rot almost every year, few wineries attempt it. Our winters aren't cold enough for icewine either, but you can, quite legally, pick your grapes, freeze them and then press off the thick sweet sludge and make quite exciting sweet wine. There are a few tasty examples – Eglantine in Lincolnshire probably did it first, but Hattingley Valley in Hampshire also excel. But this style of wine will never be more than a niche movement – something to keep the winemaker excited. Or will it? These wines sell out pretty fast at the cellar door, though it's difficult to persuade any retailers to stock them.

ORANGE WINES

I really didn't think I would be including this interesting but challenging category in a book on English wines – this style of wine originated in Georgia 8000 years ago. The wines are made from white grapes and the colour can be orange but doesn't have to be. But then I tasted Chapel Down's version in 2015 and another from Litmus Wines at Denbies. These were both remarkably refined, almost delicate, yet unmistakeably made by leaving the skins in contact with the grape juice for far longer than is usual in white winemaking. Just long enough to pick up colour, to pick up a little teasing, tannic bitterness, and a little perfume and flavour which you wouldn't normally get in a white wine – something like tamarind peel, dusty peach or apricot skins, and apple core. I should also mention the chewy, dusty-tasting Trevibban Mill Orion orange wine from Cornwall. As I said, teasing, but not overdone like many other European 'orange' wines are.

ORGANIC AND BIODYNAMIC WINES

In 2019 I had a good look at the wines from organic and biodynamic pioneers like Albury in Surrey, Davenport in Sussex and Ancre Hill in Monmouth and the fascinating, self-confident, almost exuberant wines from radical newbie Tillingham in Sussex. And to my surprise and delight, I thought – we can do organic and biodynamic in this country, too, this supposedly damp maritime nation of ours. We don't have to use all the chemicals, we don't have to protect our wines with cultured yeasts, stainless steel and sulphur. We can if we want. But there's a fascinating 'natural', non-chemical future just beginning to stretch its wings. I bet there will be more along soon.

PINK WINES

Pink, or rosé, wines will be one of the most important styles made in Britain over the next 10 to 20 years. They should be much more important now, but even in 2019 I was disappointed with the range available, and surprised at how many I tasted lacked real style. However, the standard is better now than it was, and there are more of them. In the 1990s I would do big tastings of English wines, sometimes over 100 at a time, and pink wines were regularly outnumbered by the fairly feeble and scanty reds.

That was understandable in the 1990s because pink wine wasn't fashionable, as it is now. And you can make delicious, mouthwatering pink out of your red grapes when they don't ripen fully, which is often the case in a typical English summer. Rondo, Regent and Dornfelder are varieties all making tasty pink wines and they're often better than reds made from the same vineyard. But it is Pinot Noir and Pinot Meunier which will provide the grapes for Britain's best pinks in the coming years as plantings of these two have exploded to provide the base wines for sparkling wine. These varieties both make excellent pink wine, either sparkling or still, and many vineyards will realise it makes very good business sense to release a still pink wine at nine months old rather than waiting three years to release the same base wine as a sparkler. Our cool British conditions should bring forth some of the world's most enjoyable pinks in the next few years, and I would prefer English to Provence as my rosé thirst quencher any day.

RED WINES

This section almost divides into two parts. First, there are the red wines made from grape varieties specifically bred in laboratories to be able to develop both sugar and colour in cool to positively cold environments – Rondo, Regent and Dornfelder are the most common of these – and there are quite a few tasty reds being made, sometimes with a little oak aging adding to the texture and flavour.

And second, red wines made from the great French Burgundy grape, Pinot Noir, sometimes abetted by the slightly easier to ripen Pinot Noir Précoce (early) and Pinot Meunier. Wine enthusiasts call great Burgundy the 'Holy Grail' – often sought but rarely if ever found – even in Burgundy itself. Pinot Noir doesn't like much heat, but it usually wants a bit more than is available in England, and most vineyards are better off using their Pinot for sparkling wine or for pink still wine. And yet, each year a few more wineries manage to make a red Pinot which is genuinely enjoyable to drink. It usually immediately gets showered with some kind of Burgundian comparisons, but it shouldn't. These fragile, delicate, rather sappy, occasionally floral wines with attractively bright fruit based on flavours like cranberry, redcurrant, red plum, strawberry or rosehip or pale cherry, are truly English (or, in a couple of cases, Welsh). One or two wineries such as Gusbourne, Brightwell, Bolney and Chapel Down regularly produce enjoyable Pinot reds. The 2018 vintage will see another clutch of good red wines and it won't stop there. Essex and the south-east of England have numerous vineyard sites that only require slightly more favourable conditions to produce good light red wines. And red Pinot is something that virtually every winemaker in the world wants to have a go at.

Finally, we are beginning to see plantings of warmer climate grapes such as Merlot (at Bluebell in Sussex), Cabernet Franc (at White Castle in Wales) and Pinotage (at Mannings Heath in Sussex), so the red wine world here is expanding year by year.

GRAPE VARIETIES: A RACE TO THE TOP?

The grape variety is the most important influence on how a wine tastes. All grape varieties (as with different types of fruit and vegetable) have different flavours, giving different-tasting wines. Conditions in southern England are now pretty similar to what they were in Champagne 20 years ago and so it is not surprising that the Champagne grapes of Chardonnay, Pinot Noir and Pinot Meunier can now regularly ripen well enough for sparkling wines and increasingly for still wines, too. And nearly all the new plantings are of these varieties. There are numerous early-ripening, usually Germanic, grape varieties still left in the vineyards, often making very attractive still wines and sparklers. Of these the white Bacchus and Seyval Blanc are the most important.

Ripening red grapes in the British climate is much more challenging, and erratic summer and autumn weather can still prevent proper ripening. Even so, with better varieties and techniques, results are improving rapidly. Pinot Noir and Pinot Meunier, used mainly for sparkling but also for still, in particular rosé wine, are the main red varieties but Regent, Rondo and Dornfelder are other varieties which can do well here, and are crucial for rosé and red wines in the more marginal areas.

TOTAL UK VINEYARD AREA 3579 HECTARES (2018)

Variety %	Total hectarage	Hectares planted
● Pinot Noir*	29.7%	1063 ha
○ Chardonnay	28.9%	1034 ha
● Pinot Meunier	11%	394 ha
○ Bacchus	6.9%	247 ha
○ Seyval Blanc	4.2%	150 ha
○ Pinot Gris	2.4%	70 ha
○ Reichensteiner	2.3%	66 ha
○ Madeleine Angevine	2.1%	61 ha
Others	12.5%	494 ha

Key ● Red grape variety ○ White grape variety
*includes Pinot Noir Précoce

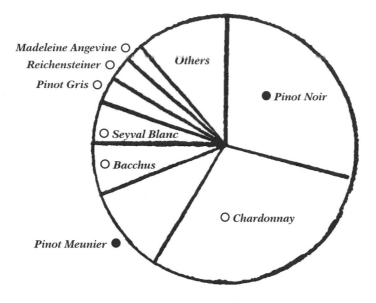

Madeleine Angevine ○
Reichensteiner ○
Pinot Gris ○

Others

● Pinot Noir

○ Seyval Blanc

○ Bacchus

○ Chardonnay

Pinot Meunier ●

WHITE GRAPE VARIETIES

CHARDONNAY

Chardonnay was planted in the very first of the revival vineyards in England – at Hambledon in Hampshire and at Oxted in Surrey in the early 1950s. Chardonnay made the greatest dry white wines in the world then – from Burgundy in northern France – just as it makes an awful lot of the best ones now from all around the globe. So you would want to be ambitious and plant the best. But it was a pity about the English climate. Year after year they could barely get the grapes to soften on the vine, let alone ripen up to a nice sugary autumn juiciness. Growers generally decided that specially bred cool climate vine varieties from Germany were a better bet, at least offering you the chance of making a wine of sorts, even if it tasted nothing like white Burgundy.

But the dream of getting Chardonnay ripe never went away and in the 1980s a few growers in Kent and Sussex dipped their toes into the Chardonnay waters again. Pretty cautiously. By 1990 only 20 hectares (50 acres) had been planted out of a national total of 929 hectares (2295 acres) of vines. Compare that with 2019 when there were 1034 hectares (2555 acres) of Chardonnay in a total of 3579 hectares (8845 acres). Yet things were about to change out of all recognition. First, Nyetimber from Sussex made a Blanc de Blancs sparkler in 1992 from 100 per cent Chardonnay grapes which won the Trophy for Best English Wine in 1997. Then Ridgeview, also from Sussex, won the Best English Wine Award in 2000 for their Cuvée Merret Bloomsbury 1996, a Chardonnay–Pinot Noir blend from Sussex and Kent grapes. These two wines simply transformed the way we looked at English wine. These were world-class sparklers and they were as good as Champagne. And the white grape variety of Champagne is Chardonnay.

I would have expected more people to rush in and plant Chardonnay, but they didn't. Did everyone think these outstanding wines were just flashes in the pan? They weren't. Ensuing vintages won trophy after trophy, and they also started beating Champagnes in blind-tasting competitions. But it wasn't until the 2003 vintage, which brought the hottest summer conditions on record to Europe and the UK, that the critical mass believing in our ability to make the world's best sparkling wine began to form. And if the UK began to believe that we could be the next great site for sparkling wines as the climate warmed up to something like the temperatures usually associated with Champagne, just a couple of hours' drive south-east from Calais, the Champagne growers began to worry that their long reign at the top of the sparkling tree might be in danger of drawing to a close. You need fresh acidity in your grapes to make lively, mouthwatering sparkling wine. Champagne growers in 2003 were complaining that their grapes were literally stewing on the vine. But take a short hop to the north across the English Channel and that weather tempered by English conditions looked like a foretaste of a golden future. At last plantings of Chardonnay began to boom, and by 2019 they accounted for nearly 30 per cent of vines.

But we hardly ever see the word Chardonnay on the front label. No, and you don't in Champagne either. We see an increasing number of wines labelled Blanc de Blancs – white wine from white grapes – and with every vintage more of these are likely to be pure Chardonnay. Some of our best sparklers are now all Chardonnay, sometimes austere, lemony and proud, sometimes stuffed with a citrus richness like lemon meringue smeared with soured cream.

Ridgeview, Harrow & Hope, Gusbourne and Nyetimber are all superb. Yet the bulk of Chardonnay grapes are used in the so-called 'Classic' sparkling wine blend with the black Pinot Noir and Pinot Meunier grapes where they definitely add a crispness, a focus, a lick of clean, clear citrus fruit to the wine.

And what about the fabled still white Chardonnay, the new Chablis, the new Meursault – is there any sign of that here? Well, yes, but the best Chardonnays don't taste like Chablis or Meursault; they taste gloriously, differently English. And there aren't many of them, yet. My first glimpse of this future was at the end of the 1980s when a Hampshire winery called Wellow produced a Chardonnay so pale it could have been water, so light and ethereal I hardly dared sip it – but the flicker of beautiful Chardonnay fruit was there, the first modern spark, barely substantial enough to be fanned into flame. The flame is now bright – not big, not furious, but bright, as vineyards like Chapel Down's Kit's Coty and Gusbourne in Kent, and Clayhill in Essex show just what can be done. Chardonnay's day job is making great sparkling wine, but you can't blame it for moonlighting now and then.

BACCHUS

So Bacchus is Britain's New Zealand's Sauvignon Blanc is it? Well, well. I suppose it could be. A generation or two ago there wasn't any New Zealand Sauvignon Blanc and New Zealand's wine reputation was far worse than England's is today. And Sauvignon Blanc wasn't exactly much sought after, either. There were a few decent examples from Bordeaux and the Loire Valley in France (where they didn't put the grape name on the label, so you probably didn't know you were drinking Sauvignon), and a few from northern Italy, and from Austria (where they helpfully call it Muskat-Silvaner). So Sauvignon came from almost nowhere to being a global star hand in hand with New Zealand becoming a global star after languishing for its entire wine existence as a grubby-tasting joke.

So there could be some sort of parallel here. Not a total one, because New Zealand swept to prominence on the back of Sauvignon Blanc's ability to make astonishing white wines the like of which the world had never seen, while its sparkling wine scene is still surprisingly unspectacular. England's rampant rush to fame is on the back of a series of sparkling wines made from the same grapes as Champagne, in the same way as Champagne, and often from very similar soil and climate conditions, which have equalled and increasingly surpassed the quality of equivalent Champagnes. But the still table wines, of any colour, lag well behind.

Frankly, there is no dishonour in this – Champagne makes very few still wines you'd want to drink, and Burgundy, using the same grape varieties but grown further south, doesn't make many exciting sparklers. Most famous regions specialise. Well, yes and no. Most famous European wine regions specialise, sometimes literally only producing a single type of wine from a single grape variety. But most New World wine countries are keen to try as many different wine styles as they think might work. Europe usually has rigid wine laws saying what you can and can't do, often worked out over the centuries. The New World is ready to give everything a go. And with England at the very northern tip of being able to barely ripen grapes like Chardonnay and Pinot Noir, which always make the best fizz when they're barely ripe, but not the best still wine, another grape variety is going to be needed to spearhead any kind of still wine revolution.

Because England is so cold in wine terms, not many top grape varieties will ripen here. Producers such as Rathfinny and Denbies have tried the cool climate classic Riesling without success. Chenin Blanc from the Loire Valley hasn't worked, either. You would think that Sauvignon Blanc might work. Well, it may, since tasty examples are beginning to creep out from wineries such as Woodchester in the Cotswolds, Denbies and Greyfriars in Surrey, Polgoon in Cornwall and Sixteen Ridges in Herefordshire (the polytunnels help with these two). Even Albariño (the grape from Galicia in the cool north-west of Spain) has made an appearance at Chapel Down. No, it needed one of the specially bred German vine crosses to step up – preferably one without too long a name and too many umlauts. Bacchus fits the bill.

Bacchus was 'created' in Germany as long ago as 1933 and, like most of these vine crosses, is pretty much a sugar machine in Germany's warmer conditions. But in Britain, Bacchus ripens much more slowly and holds on to its acidity and consequently its bright elderflower and hedgerow scent that would be lost in sunnier climes. The flavour can be a little sappy, the acid reminiscent of mild lemon zest, and the fruit might even be peachy or at very least as ripe as a good English eating apple or a crisp greengage plum. The wine can be dry, or not quite dry. One or two producers like Chapel Down have even tried aging it in oak barrels and to my surprise it works quite well. There is even some sparkling Bacchus. It took a while for Bacchus to get going here – I did two tastings of over 100 English white wines in 1991 and there were only three Bacchus wines included, and only the wine from Three Choirs was much fun. It's a different story now. Obviously wine judges like Bacchus because its wines regularly win the top still white wine awards in wine competitions in the UK, although sunny vintages like 2018 will probably give Chardonnay a healthy kick up the charts. And Bacchus will grow just about anywhere. It's excellent in Essex and Norfolk (New Hall, Flint and Winbirri), excellent in Kent and Sussex (Chapel Down and Bolney), and is happy right across the West Country down to Cornwall (Camel Valley). And the British wine-drinking public are catching on, too – just as they did with New Zealand Sauvignon Blanc. The flavour's not that similar, but the wine is fresh and bright and fairly low in alcohol – and you couldn't get a better name for a wine grape than Bacchus, the Roman god of revelry and wine, the god of fun.

SEYVAL BLANC

Every dog has his day. Every grape variety will find somewhere in the world where it can flourish. Even vines that are banned, uprooted, exterminated and despised can still find some little corner, some little nook, where soil and climate just briefly intersect with a grower's passion and talent. And a super wine can result. Seyval Blanc was created through cross-breeding in France in 1921. But it's banned from all the French quality wine areas because it isn't a pure *Vitis vinifera* species (the grape species from which nearly all the world's quality wine is made), and they're very strict about that in France.

Actually it does make pretty rubbish wine in France, but in England, it finds its niche. It ripens early, so can cope with the vagaries of the English climate. It's got high acids, and doesn't have a particularly appealing flavour by itself. There's something slightly feral, slightly like picnic nuts that had fallen in the mud, but you ate them anyway, like eating green apple peel by itself without the flesh – and did you wash your hands? Owen Elias, once of Chapel Down, and one of England's top winemakers, says Seyval Blanc tastes of raw potato and cabbage.

So you sort of wonder why anyone should plant it? Well, it absolutely divides opinion, but I've been tasting Seyval Blanc for over 30 years; I remember some spicy, vivid, very appetising Seyval from Breaky Bottom made in years like 1984. Spots Farm at Tenterden in Kent made green and smoky but honeyed dry Seyval in 1983. And these – especially the wine from Breaky Bottom – aged for years, and really did develop flavours more like dry French wines than those of any other variety. Which was probably the whole point. Nearly all the grape varieties in the early revival after the Second World War were German-inspired and made off-dry, rather scented whites at their best (some of the original winemakers tried to make them dry, but few outside the dental profession found much pleasure in them). Seyval could give a passingly decent impersonation of something like a Muscadet or a Petit Chablis. To be honest, in the hands of someone like George Bowden at Leventhorpe Vineyard in Yorkshire, it still does.

But Seyval in southern England is now there largely because it can produce good crops of fairly neutral, quite high acid juice that works well as the basis for sparkling wine or as part of a sparkling blend. None of the top sparkling blends include Seyval any more, yet producers like Camel Valley, Breaky Bottom, Bluebell, Denbies and Three Choirs still include it in some of their blends. You can still just about taste that rustic green apple and muddy nut quality, but I'm really quite fond of it. After all, it's very English, and no one else does it quite like that.

PINOT BLANC AND PINOT GRIS

Since these two white and off-white members of the Pinot family are reckoned to ripen quicker than Chardonnay you would expect a rush to plant them in England and Wales. Well, there hasn't exactly been a rush, but they are slowly making their presence felt, and since they are both permitted varieties in Champagne, they do have a certain attraction for English bubbly producers.

Pinot Blanc is often rather disingenuously called 'Chardonnay' around the world since Chardonnay is an easier name to sell. The fact that Pinot Blanc ripens more quickly and gives bigger yields than Chardonnay makes the deception profitable. There is no need for deception here in the UK since we are at the beginning of the journey toward fine dry white still wines. And Pinot Blanc works well here. Chapel Down has made a really nice fresh Pinot Blanc since the turn of the century and Stopham in Sussex have specialised in it. Stopham also makes a full, spicy, slightly tropical Pinot Gris, as do Bolney (also in Sussex) and others. This is the same grape as the Italian Pinot Grigio – which is usually a fairly neutral wine. But Pinot Gris is the French name for the grape and in Alsace it can make succulent, honeyed wine. This is the path English producers will be following. Some producers like Rathfinny blend the two together.

MÜLLER-THURGAU

During the 1980s the British drank more wine from the Müller-Thurgau grape than from any other variety. Why? Because most of the wine we drank then was pretty ordinary German stuff like Liebfraumilch or Piesporter Michelsberg – and Müller-Thurgau, easy to ripen, and likely to literally weigh itself down with vast crops of tasteless grapes, was the workhorse of such wines. But shove Müller-Thurgau in the damper, colder soils of Britain, where yields of grapes were nothing like so gross as in Germany, and where the grape only crept to ripeness even in warm years – and you could make very nice wine from this much abused grape. In the 1990s it was England's most widely planted variety, as English producers tried to ape the Germans. At its best, from producers such as Pulham St Mary in Norfolk, Breaky Bottom in Sussex, Syndale Valley in Kent and Wootton in Somerset, Müller-Thurgau was lightly floral, with a pleasantly crisp leafy green acidity and usually a few grams of sugar left in it. Better than Liebfraumilch? Definitely. But there is now barely a vine left in England.

OTHER WHITE VARIETIES

Although Chardonnay and Bacchus are the two grape varieties that get the most attention, and simply by themselves cover about 40 per cent of the British vineyards, there are various other grape varieties that were famous once. Seyval Blanc and Müller-Thurgau are two obvious examples, and there are a few varieties, carefully bred for the colder, damper conditions of many parts of Wales and the north of England, which deserve a mention.

In the 1980s and '90s, you'd see an awful lot of wines labelled Müller-Thurgau. But there were various other Germanic vine crosses too, some of which were rather good. Huxelrebe doesn't sound very appetising but can make very snappy, leafy, rather gooseberryish wines, especially in Kent (I always liked Biddenden's Hux), and there's still some planted. Siegerrebe is a very early ripener that can make rich, honeyed, almost tropical wines – Three Choirs in Gloucestershire are famous for it. Ortega has turned out to have a genuine role to play – it ripens fast and is very susceptible to noble rot, the fungus which concentrates the juice and gives enough sweetness to make dessert wines. Denbies regularly makes a cracker in Surrey. Varieties like Reichensteiner, a rather bland but early ripening variety, Schönburger, Würzer, Würzburger and Ehrenfelser are quietly fading away. Madeleine Angevine – Mad Ange as growers call it – is proving surprisingly resilient and makes very attractive wines in Suffolk (Giffords Hall), Hampshire (Danebury), Yorkshire (Leventhorpe) and elsewhere – mellow, soft, ever so slightly smoky. But there is a bunch of new varieties, very specifically bred for the cold – they're popular in countries like Denmark and Sweden. The most important are Solaris, Orion and Phoenix. In Wales and the north of England, they are crucial and produce wines with real flavour and personality. If global warming continues its relentless march, they may eventually give way to the big beasts like Chardonnay, so enjoy them for what they are – the best varieties at the moment for some of our marginal areas.

RED GRAPE VARIETIES

PINOT NOIR

You simply can't stop winemakers planting Pinot Noir. All over the world, from the heat of the eastern Mediterranean, to Australia and Argentina, to California and the Pacific Northwest, as well as all over western Europe, wine growers crave success with Pinot Noir and so Pinot Noir is planted, rarely with memorable results, and usually in conditions that are too warm and too dry for it to flourish. Burgundy, its heartland in eastern France, isn't warm and isn't dry and Pinot Noir only ripens satisfactorily on a tiny sliver of one east-facing slope. Champagne, further north in France, is less warm and less dry, and despite rarely making still red wine from Pinot Noir that would turn many heads, the grapes do ripen enough to form the basis for superb sparkling wine, since sparklers are best made from grapes that are barely ripe.

And then there's Britain. Above all, the south of England, stretching from Kent and Sussex to Hampshire and Dorset via Surrey and the Thames Valley. Here, it's even cooler than in Champagne. Here, it's often even wetter. But as the climate changes, particularly in the last 20 years, more and more vineyard sites are proving that they can manage the balancing act of slowly edging the Pinot Noir toward ripeness, usually in October, without ever letting its delicate, almost fragile cool climate flavours descend into the jammy ripeness which occurs so readily in warmer climes. Such grapes, with their long period sitting on the vine as the summer fades and the mellow warmth of autumn takes over, can build up fabulous flavours at low alcohol levels – perfect for sparkling wine.

And Pinot Noir is massively helped in the south-east and in Hampshire by soils – often chalky, sometimes sandy – that can be indistinguishable from the soils of Champagne, which is, after all, only a couple of hours' drive from the Channel port of Calais. These British soils have shown that they can produce grapes as suitable for sparkling wine as the soils in Champagne. Yet the fascinating thing about Pinot Noir in England is that it also excels on heavy clay soils, like those of the Kentish Weald or the Essex coast. It doesn't excel on clay in France. In France they will tell you that Pinot Noir is a very fussy grape and the marriage of soils, aspect to the sun, amount of heat, sunshine hours, the wind, the rain – all that – must just fit precisely together for this variety to produce interesting results. That's a lot of variables, and even Burgundy and Champagne struggle to piece the jigsaw together.

Will Britain find it any easier? No. But has Britain got conditions as good as those in France for Pinot Noir? Well, for still red wine, maybe not. Yet since cooler parts of the world have at last been able to produce good to excellent red wine that really doesn't taste like

Burgundy at all, England might find some special places where an entirely English-tasting red wine is possible. And it won't be from conditions or soils like those of Burgundy. Pinot Noir doesn't like heat, but it rots easily and so likes to stay dry. Kent and Essex are two of the driest parts of England. Even on heavy clay soils, Pinot is beginning to make wines to turn a head or two from these counties. Not like Burgundy, but nice.

As the climate warms, we will certainly see more wineries making a stab at Pinot Noir red, but right now, its best role as a still wine is pink. Less ripe grapes with less colour and sugar can make delicious, refreshing pink wine. Particularly after the explosion in planting during 2017–19, there are going to be a lot of young Pinot Noir vines producing grapes that need to be quickly turned into wine and sold for cash flow reasons alone. Pinot Noir pink is perfect for this, and we should shortly be making some of Europe's best dry rosé.

But Pinot's main job is to provide the base wine for sparkling wines. Pinot Noir is a black grape, but the juice is colourless, and if you press the grapes very carefully you can avoid staining the juice. Two-thirds of the grapes in Champagne are black, and most of the fizz they produce is white. Champagne employs the most delicate, precise grape presses in the world, and these presses are now appearing all over England, demonstrating just how serious the winemakers are. There are now 1063 hectares (2626 acres) planted with Pinot Noir in England and Wales, out of a total of 3579 hectares (8845 acres). That makes Pinot Noir the most planted variety by a whisker from Chardonnay – each year, Pinot Noir is jostling with Chardonnay to boast the highest acreage. And despite being less predictable when it comes to things like setting a crop, grape growers are surprised at how adaptable it can be if you care for it in the vineyard. Clearly it revels in the chalk and sandstone and gravel soils, but its ability to give really tasty crops from clay, so long as the weather is dry, is something of a revelation.

The majority of the Pinot Noir wine is blended with Chardonnay to make what is called a 'Classic Cuvée'– the fuller, broader textures and flavours of Pinot Noir (and sometimes, Pinot Meunier) complementing the leaner, more vivacious character of the Chardonnay. White Blanc de Noirs from Pinot Noir by itself is so far a less popular style but is bursting with flavour and personality and wineries like Exton Park, Ridgeview and Jenkyn Place make wonderful examples. Most of the pink sparklers are made by blending Pinot with Chardonnay, but there are some so-called Rosé de Noirs pinks that match the best from Champagne, from wineries like Coates & Seely, Hattingley Valley, Ridgeview and Exton Park.

Pinot Noir may be finicky, may be mercurial, may be unpredictable, but in Britain, for sparkling wines above all, it has found a home.

Pinot Précoce (early Pinot) is a mutation of Pinot Noir that ripens earlier and gives wines of body and colour though less perfume and beauty. Précoce is useful for adding a little colour to Pinot Noir, and is very useful at the margins of British grape-growing where the Pinot Noir simply wouldn't ripen. It is also known as Frühburgunder.

PINOT MEUNIER

This is an interesting variety, particularly because it might actually have been in England far longer than any other black grape variety. Meunier means 'miller', and the leaves do look as though they've been dusted with flour. Meunier is widely grown in Champagne for sparkling wine, but hardly figures when it comes to still wine anywhere in the world. Yet one of the early pioneers of modern English wine, Edward Hyams, found a vine growing at Wrotham in Kent in the 1950s which he thought might be Pinot Meunier, and he named it Wrotham Pinot. By all accounts it ripened early and gave grapes with higher sugar levels than Pinot Meunier vines obtained from Champagne, but it seems to have disappeared without trace. A pity. I once had an example from Devizes in Wiltshire which was unusual, honeyed and excellent. I wonder if someone has a vine or two in their back garden that we could propagate, because in England's climate any classic French grape with an early ripening habit would be welcome. In fact, a lot of the sparkling wine producers have now planted Pinot Meunier and are mostly pretty happy with it. Some, like Nyetimber and Exton Park, have made fabulously tasty single varietal sparklers from it, while Simpsons in Kent has made a smashing still dry rosé.

DORNFELDER

I really thought I had seen the future when I tasted Denbies' Dornfelder 1990 red. A dark red colour, a fabulously ripe, strong, damson and cream aroma and then a flavour which was fairly high in acid, yet bursting with cherry and damson fruit, wreathed in a warm shroud of smoky spice. Côtes du Rhône? Burgundy? I wrote, 'This is brilliant in English terms and very good in any terms.' And I can remember it so vividly, because I'm still waiting for the next superb English Dornfelder. I haven't tasted it yet. Which is a pity. Dornfelder is one of Germany's most successful red grape inventions. It may take a fair while and a few more years of climate change before Pinot Noir can regularly hit decent ripeness for making still red wines. Dornfelder can already achieve it. But very few are giving it a go. Perhaps if it had a French name it would stand a better chance.

RONDO

I don't have a particularly good relationship with Rondo, although I am slowly coming round to it. Obviously any grape variety that can produce a dark, dense red wine in Britain is going to be enthusiastically planted by numerous growers, and I get that. It's just that I don't think simply looking dark and brooding in the glass is good enough. I want it to taste like something I'll enjoy drinking. And Rondo doesn't find that easy. Well, this is probably something to do with Rondo's parentage.

Rondo is a hybrid with a decent chunk of parentage coming from the River Amur on the Chinese–Siberian border. They don't get much summer there, so the grapes ripen quickly, and Rondo's red flesh and juice gives you every chance of a darker red wine than you'd have thought possible in England – or Wales, since one of the best examples comes from White Castle near Abergavenny. The fruit spectrum is at the bitter end – sloes or damsons, chewy, rather coarsely meaty, like shredded chuck steak. Often that's because people have tried so hard to prove with Rondo that they can make a serious English red that the thing ends up like cloddish soup. Recent successes show winemakers being much more careful and delicate, using less oak, and sometimes blending in lighter varieties. The result is a much pleasanter drink, and still quite full-bodied enough to score well on the testosterone counter.

OTHER RED VARIETIES

Red and pink wines are certain to assume greater importance as
the climate warms. But the emphasis is sure to be on Pinot Noir,
which, with its numerous clonal variations, will occupy the dreams
and ambitions of most of Britain's red winemakers. Even so, during
this period when few producers manage to ripen their Pinot Noir
sufficiently for a satisfying red wine, other varieties will be needed
for a while yet. Pinot Noir Précoce, also known as Frühburgunder, is
showing its quality already. This mutation of Pinot Noir ripens a little
earlier, has a darker colour and a decent earthy flavour sprinkled
with cherry and plum juice. It's very useful to give a bit of colour
to Pinot Noir, but is good on its own. Regent is a German crossing
with the tasty Chambourcin in its parentage. Much gentler and
mellower than Rondo, it can even achieve slightly strawberry Pinot
quality. Again, good on its own, it's useful for blending with Rondo
and calming the beast down. There's some South African Pinotage
in Sussex at Mannings Heath, and some Bordeaux Cabernet Franc
at Yorkshire Heart and White Castle in Wales. There's even some
Cabernet Sauvignon and Merlot at Sharpham in Devon and Sixteen
Ridges in Herefordshire. They're grown under plastic for now, but
good things will come to those who wait. Bluebell are already using
outdoor-grown Merlot in a very tasty rosé blend.

Of course, if the scientists had their way, all this discussion of
which red variety might work best would be theoretical. The East
Malling Research Station in Kent has a worldwide reputation
for its research work into fruit, above all into apples. Now the
scientists have turned their attention to grape varieties, and think
they have come up with the perfect red wine variety for Britain –
the high-yielding, early ripening, disease-resistant Divico. This is
a variety originally bred in Switzerland – not a bad start for a cool
climate variety – and I'm sure that British vineyards will give it
a try (Halfpenny Green has already made some). But when East
Malling's lead scientist says, 'Divico is likely to supersede Pinot
Noir', I think that's a scientist's view. Try asking a winemaker what
challenge excites you most? Making great Pinot Noir or making
great Divico? I know the answer, and it isn't Divico. Even so, if it
can make good, ripe, affordable red wines better than Regent or
Rondo, then it will be a success. But the final arbiter of success will
be us, the wine drinkers. Do we actually like the taste?

BRITISH BUBBLES

Good bubbles lift your spirits just by gazing at them in the glass – isn't it the bubbles in the wine that are important rather than the flavour? Well, creating good bubbles requires skill, patience and money but the flavour and acidity of the grapes really matters. All around the world winemakers have tried to match Champagne but have failed because they simply can't grow grapes with the same characteristics; England, however, looks as though it can.

WHAT IS NEEDED TO MAKE GOOD FIZZ

In a way, sparkling wine is one of the most industrial of all wines. It relies completely on human intervention, and there are a whole series of mechanical activities that need to take place if you are to end up with a delightful, limpid, foaming glass of guaranteed happiness at the end of it all. And if it's such an interventionist wine, reliant on the chemist and the technician, does the exact type of grape juice matter, or can't the chemist adjust it according to what he requires? Does the grape variety matter when the juice is going to be so played about with? And surely the precise vineyard doesn't really matter. Does it?

Well, luckily it does. The grape variety and the climate it grows in and the soil of the vineyard are of supreme importance. And, interestingly, being *un*able to fully ripen your grapes is a matter of massive importance, because the natural high acidity in such grapes is crucial for the vivacity, the liveliness, the excitement and the thrill that a glass of fine fizz gives.

CHAMPAGNE, THE ORIGINAL FIZZ

Champagne has been thought of as the world's top sparkling wine for 300 years and more. All round the world people have striven to recreate the flavours and the textures of Champagne – and they have failed. Even using precisely the same grape varieties, the same methods of production, the same machinery – they have failed. Because the one thing other countries have failed to reproduce is the actual taste and character of the Champagne grapes themselves. And for that, you needed the windy, rainy, stormy vineyards huddled around the Marne Valley to the north-east of Paris, where there wasn't enough warmth to ripen grapes enough to make a decent still wine, red or white, but where, as the pale summer sun slipped toward a damp, chilly autumn, Chardonnay and Pinot were just ripe enough, had just enough sugar, to ferment into a thin, lean acid wine which human ingenuity would transform into a cascade of golden bubbles and sumptuous flavours. Nowhere else seemed to be able to provide such marginal grapes. Until now. Champagne's nearest wine neighbour shares the same soils, the same grape varieties, and now, almost the same weather. England. Champagne is getting warmer, perhaps too warm. England is getting just warm enough.

It's quite probable that England was the first ever major market for sparkling wine. It certainly wasn't France. The Champagne wines were very popular at the French court of Versailles in the 17th century, where Louis XIV, the Sun King, thought they were good for his health. He did live an awfully long time, and he was on the throne for 65 years. But he hated bubbles in his wine, and the chief job of the winemakers supplying the court was to try to remove any bubbles in the wines before they got to his royal lips.

This wasn't easy. Vintages were generally late on the chilly slopes around the Marne Valley. And winter closed in very quickly, often before the wines had finished fermenting. The cold meant that the yeasts couldn't continue working. So you usually had a pile of barrels full of wine that hadn't finished fermenting, and if you tried to sell these to the king of France he'd have your *tête* on a plate.

SO DID THE ENGLISH INVENT CHAMPAGNE?

Help was at hand. Good old England. Each spring these barrels would be shipped to London, unfinished. As the weather warmed up, the wine started fermenting again and for a few weeks the London pleasure gardens were awash with sparkling wine – or rather still-fermenting wine. Everyone had a lot of laughs and monumental hangovers. Just for a few weeks? This was far too much fun for it to merely last a few weeks in spring. Luckily England was the one place in the world which was fitted out to prolong the pleasure. England had adopted coal as a fuel before any other nation – chiefly because the country was running out of trees. Coal burns far hotter than wood, so glass made in a coal-fired foundry is far stronger than traditional European glass. The French rather enviously called this new glass, which was so much stronger than their delicate stuff, *le verre anglais*.

The cidermakers of Gloucestershire were early adopters of this strong glass. They put dry cider into the bottle, added a couple of raisins and a walnut-sized lump of sugar, banged in a cork and the whole thing began re-fermenting in this strong glass bottle which could cope with the pressure – and the stopper could also cope because it was cork. The rest of Europe had the lost the knowledge of cork as a brilliant hermetic seal. England, friendly with Portugal, the main producer of cork, since the Treaty of Windsor in 1386, was the only place that still knew the secret. Sparkling cider became known as 'the English Champagne' in London society.

It took a man called Christopher Merret, an English physician and scientist and Fellow of the Royal Society, to demonstrate in 1662 that what worked for cider could work for wine: strong English glass bottles filled with wine that had been shipped in barrel from Champagne, add a little sugar, and then push in tight corks tied down with twine – and wait. Wait for six months, one year, two years – the wine would have re-fermented inside its bottle, and when you pulled the cork, there would be an explosion of foam and froth and fun – all year round, not just for a few head-banging weeks in the spring.

So does all this mean you can say the English were the ones who invented Champagne? Well, sort of yes and no. Dom Pérignon, the French winemaker monk who is supposed to have invented the method of a second fermentation in the bottle, creating bubbles in 1698, was one of the chief people trying to get *rid* of the bubbles. But he did respect those hardy English bottles, and in the 18th century he and others slowly began to turn Champagne from being a still wine into a bubbly one. After which Champagne never looked back and by the 19th century had become world famous as the finest fizz ever created.

And England? It's all very well taking barrels of French wine and making it sparkle. But what about English wine? Wasn't there a rush of interest? It doesn't look like it. England was in the middle of a little ice age in the 17th and 18th centuries. Planting vineyards wasn't uppermost in people's minds. A few vineyards were established in places like Surrey, and some did produce some sparkling wine, but none of the vineyards lasted long. One guy called Charles Hamilton at Painshill made what he thought was pretty good fizz, but he found it difficult to persuade many people to share his view – 'Such is the prejudice of most people against anything of English growth.' It wasn't until the 21st century, our own century, and probably not until its second decade, that such opinions of our home-grown wines began to fade.

BUBBLES - HOW THEY'RE CREATED

So what's the secret to making great sparkling wine? Well, despite all the gadgetry which I'll talk about in a minute, you simply must have the right grapes. All over the world people have tried to make Champagne-like fizz but are let down because they can't grow grapes with the right flavour. Ideally they should be the Champagne varieties of Chardonnay, Pinot Noir and Pinot Meunier, although Seyval Blanc makes a pretty good, rather individual base wine, too. They should be planted somewhere with a cool climate that barely allows you to ripen them, preferably on chalk or sandy clay soils, although gravels and even heavy clays have shown perfectly good results. And if you want somewhere that fits this bill, southern England – especially on the chalky North and South Downs and the Thames Valley and on the greensands and sandy clays that crop up a fair bit in Kent and Sussex – is ideal. From a virtual standing start with Nyetimber in the 1990s, over 70 per cent of English grapes are now the Champagne varieties.

Fine. That's nature. What about the interventionist human hand? Well, right from the beginning, nature is being manipulated. In Champagne, two-thirds of the grapes are black, and well over half of the English ones are too. Yet they mostly make white wine. The crucial thing here is that the juice inside the black grapes is colourless, so if you press them very, very gently, you can coax out half, maybe more of the juice before the squashed skins begin to stain them pink and red. The best sparkling wine presses are shallow and shaped like a big circular disc. Other good ones are more rectangular but use a great inflatable cushion to slowly massage out the juice. The best wineries have one of these two.

Gentleness is everything. If you squeeze too hard you start getting sappy tastes, resinous tastes, chewy textures and all the delicacy you'd hoped for is gone. But these later 'pressings' of the grapes can have lots of flavour and are often used as the basis for really tasty still white wine blends. Chapel Down make their excellent Flint Dry wine by using Chardonnay pressings mixed with whatever else they've got.

So you've got your juice; any heavy sediment has been removed. Let's go. Time to ferment. Usually this is done in a stainless steel tank where you can control the temperature. Some people use a few oak barrels. There's even a concrete egg or two and an earthenware Georgian *kvevri*. But mostly stainless steel is best, because you generally want a purity of flavour. Fermentation in oak barrels will add richness to the wine – some of the more expensive wines benefit from this. There are highly efficient cultivated yeasts on the market – the same ones as used in Champagne – that guarantee a quick, problem-free fermentation. That's what most fizz makers want. So most wineries use these cultured yeasts. Some let the numerous yeasts that live in the vineyard and in the winery (wild yeasts) do the fermentation work, saying it's more natural. Maybe it is. Maybe it can give wine with more character. But maybe it's less consistent and more likely to add wild flavours to the wine. Natural yeast or cultured? It's neither better nor worse, but in sparkling wine, most producers prefer the cultured yeast.

The fermentation is finished. You now have wine, but with no bubbles. So what next? In England the acid level of the wine is often fairly high and the sharpest effect is from malic acid. This is a green raw acid that can be delightfully mouthwatering in young table wine like Bacchus, but in sparkling wine the less assertive tartaric acid is the one most winemakers like. So they will frequently reduce the malic acid by inducing a bacterial conversion of malic acid to the much softer, creamier lactic acid. This is called malolactic fermentation, but it isn't really a fermentation – it doesn't create any alcohol – it's a bacterial conversion.

You may want to put some of your wine into barrels for a few months to soften it up. And you'll have various vats and tanks, big and small, with different batches of wine from different grape varieties, from different blocks of vineyard, or from different picking dates. You might be making a wine from a single harvest – a vintage wine. You might be making a non-vintage or multi-vintage blend of the wine of several different harvests. You might be trying to separate off your very best batches into a 'prestige cuvée'. And obviously you may have decided to make some pink wine. But whatever your ambitions you'll have to draw samples from all the vats and barrels, put them up on the tasting table and try to blend together a liquid that will be your ideal sparkling wine in two to five years' time.

If this sounds like fun – well, it's very rarified fun. The new wines are young, acid, unformed; and there might be so-called 'reserve' wines from several older vintages as well to consider for your non-vintage blends. But winemakers often see this blending, which frequently takes place in the icy dog days of winter, as one of the great challenges – and great joys – of their year. I've tried it a few times and it's tough, trying to see into the future, when an uplifting balanced golden wine cascading with bubbles and laughter will flow into the glass, yet all these samples are raw, rough to taste – and they still have no bubbles.

SO NOW'S THE TIME FOR BUBBLES

The classic method of making wine sparkle is by creating a second fermentation in the bottle of this dry still wine. It won't start fermenting again by itself – you'll have to encourage it. So you put together what the French call a *liqueur de tirage*, basically a mixture of a precise amount of sugar and yeast; you then add this to the still wine blend in the tank – give that a really good mix to integrate the sugar and yeast (you could add the yeast mix direct to the bottle, but it's usually done in the vat) – and then you bottle this brew, snap on a crown cap and you wait.

You've added yeast and sugar to wine. Scientifically, it has to start a new fermentation. This will produce about one degree more alcohol – and vast amounts of carbon dioxide. But your wine is in a hermetically sealed bottle. Where can the gas go? Luckily, carbon dioxide is a very soluble gas so the whole lot dissolves in the wine and creates immense pressure. That was why the invention of really strong glass for bottles in England way back in the 17th century was so important for the birth of a sparkling wine world. They could withstand the pressure.

Ideally this second fermentation takes place slowly in cool conditions, with the bottles on their sides. It may well take a month or two, and as the yeast cells finish their job of fermenting, they die and drop to the bottom of the wine, which, with the bottles horizontal, is along the side of the bottle. This creates a sticky sludge (though some modern yeasts aren't sticky), and this sludge has a fantastically important part to play. As those dead yeast cells gradually decompose they undergo an enzymatic reaction called autolysis which releases all the flavours and chemical components in the yeast into the wine.

This is crucial for quality. Since an English sparkling wine is sure to be made from acid base wine, this 'autolysis' is the natural way to let the wine develop the flavours of cream and roasted nuts, brioche and croissant crust that are the joy of fine sparkling wine. All those flavours are in the yeasts, but they need time to seep into the wine. Most experts reckon the wine must lie there for at least 18 months for this gastronomic decomposition to occur. Top producers often leave their wine on the lees sediment for three, four or five years. If you whip a wine off its yeast lees before they have begun to decompose, you lose the chance of any richness in your wine. This 'autolysis' transforms an acid sparkling wine into a mouthful of delight.

But it won't look that appetising if the glass is full of the sludge of decomposed yeasts. Time for some more human ingenuity, invented, obviously, by the Champagne producers. You have to dislodge that yeast sludge, get it down to the mouth of the bottle, and then somehow remove it without losing half a bottle of foaming wine at every attempt. In the 19th century the Champagne companies invented a method of gradually taking each bottle from horizontal to standing on its head over a matter of months. Armies of men would methodically work through the vast cellars, upending the bottles bit by bit and giving them a gentle knock as they did so to dislodge the yeast. At the end of the process, called *remuage* or riddling, all the bottles would be standing on their heads with a lump of sludge sitting in the neck above the cork. You can still do this process by hand, obviously, but producers nowadays employ machines called gyropalettes. These are cages containing up to 500 bottles which are computer-controlled and which methodically shake and upend the bottles over a matter of days rather than months. Not so romantic, but not so many workers calling in sick with arthritis of the wrist.

But we're not finished. We've got a cellar full of bottles standing on their heads with the necks blocked with sludge. Ah, another French invention – disgorgement. With a lot of practice, you could probably learn how to flip off the crown cap and turn the bottle upright all in one, and just lose the plug of sludge, not loads of wine. But there's now a machine to do this. The bottle necks are submerged in freezing glycol solution, then upended mechanically with almost no loss of wine except for the frozen plug of dead yeast cells. So at last you have bone-dry sparkling wine.

You can leave it bone-dry, but very few producers do because it can feel a bit abrasive and lack charm. So usually a little dosage of wine, sugar and sometimes a dash of brandy are added, which lets you imperceptibly sweeten the wine. If you do want your fizz to taste slightly sweet – and several wineries like Nyetimber, Hattingley Valley and Ambriel do make a bit of slightly sweet demi-sec fizz – well, you add a bit more sugar. And then you bang a Champagne cork in, tie it down with a wire cage and you're done. You should probably age the wine for a few months to let it settle down after all this palaver. You can age it for several years if you like. But you've now made yourself a beautiful bottle of traditional method English sparkling wine.

DISGORGEMENT DATES

If you look on the back label of a bottle of English sparkling wine, some now include the date the wine was first put into bottle to start its second fermentation and the date of disgorgement. (This was first done in Champagne in 1985 by Bruno Paillard.) The second date shows how long the wine has rested on its lees, or yeast cells. A longer time usually means a richer, more complete, more rounded wine.

A TOUR OF THE REGIONS

If you look at the map on page 23 you will see that the biggest concentration of vineyards is in the south-east of England, with Kent and Sussex dominant. Further west Hampshire, part of the Wesex region, is also important. This is because the soils, especially the chalk soils of the North and South Downs, are ideally suited for vineyards, and the weather is warmer and drier than further west.

Even so, there are good vineyards in Dorset, Devon and Cornwall, and good wines are produced in the west as far north as Worcestershire. But conditions are more challenging in Wales, the Midlands and the north, even though climate change is encouraging more and more people to have a go. There's a scattering of vineyards in East Anglia where Essex, especially, has some of England's driest conditions and some of the best potential for grape-growing in the whole country.

DISTRIBUTION OF PLANTINGS BY WINE REGION (2018)

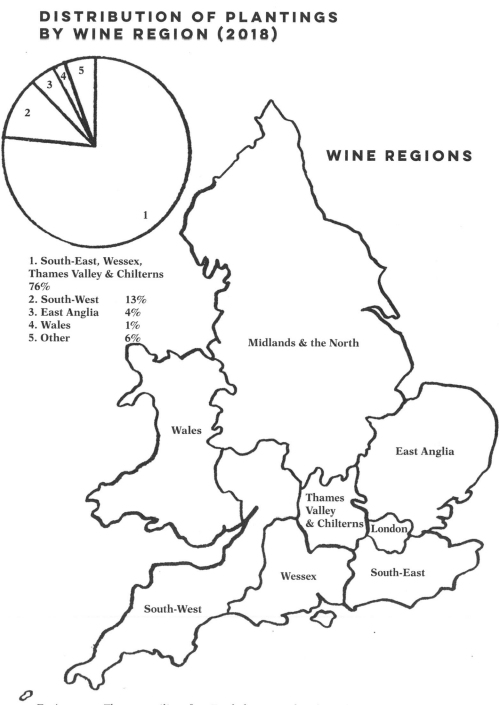

1. South-East, Wessex,
Thames Valley & Chilterns
76%
2. South-West   13%
3. East Anglia   4%
4. Wales   1%
5. Other   6%

WINE REGIONS

Midlands & the North

Wales

East Anglia

Thames
Valley
& Chilterns

London

Wessex

South-East

South-West

Facing page: The tranquility of an English vineyard at the end of a glorious summer – this is 2018 vintage time at Albury in the Surrey Hills.

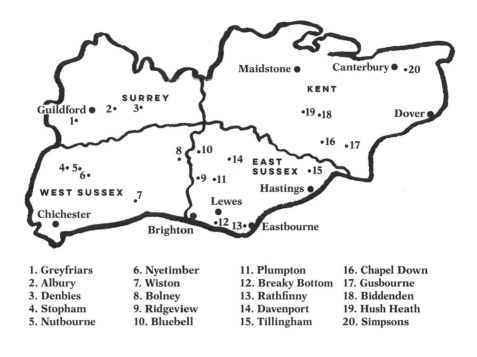

1. Greyfriars	6. Nyetimber	11. Plumpton	16. Chapel Down
2. Albury	7. Wiston	12. Breaky Bottom	17. Gusbourne
3. Denbies	8. Bolney	13. Rathfinny	18. Biddenden
4. Stopham	9. Ridgeview	14. Davenport	19. Hush Heath
5. Nutbourne	10. Bluebell	15. Tillingham	20. Simpsons

SOUTH-EAST

This is the engine room of the great English wine revolution that is hurtling ahead at the moment. Both Sussex and Kent could claim leadership, but sharing the plaudits is a better way forward. Most of Kent's vineyards, until the 21st century, didn't make use of her greatest asset – the billowing slopes of south-facing chalk soils that follow the North Downs across the county from the cliffs at Dover. Most of the early vineyards were either planted on sandy clays – good vineyard soil – or on heavy Wealden clay, which is a pretty tough medium to grow anything on. Most of those early vineyards are gone, but some including Biddenden still flourish. And new ones like Hush Heath show that you don't have to be planted on chalk to excel. Even so, the most exciting events are now taking shape along the North Downs, from Simpsons in the Elham Valley near Canterbury to the remarkable Chapel Down developments based on Kit's Coty near Maidstone and across to Squerryes near Sevenoaks. The biggest new investment in vineyards by the Monaco millionaire Mark Dixon is in Kent, and the Champagne house Taittinger chose Kent to plant their first English vineyard – Domaine Evremond at Chilham near Canterbury.

Sussex is technically two counties – East and West – and they are both equally important. East Sussex saw more of the early development by estates such as Carr Taylor, Breaky Bottom, Davenport and Ridgeview, is home to Plumpton College and its wine centre, and has seen renewed investment by estates like Rathfinny, whose championing of Sussex has led to an application for a Sussex 'Protected Designation of Origin' or PDO. This is by no means supported by all the Sussex producers. West Sussex is most famously the home of Nyetimber, but has also seen major developments by estates such as Bolney, north of Brighton, Wiston, near Worthing, and Tinwood near Chichester. The South Downs run through the whole of Sussex, but often as coastal cliffs and uplands, and most of the successful vineyards have been established on sandy clays and the greensands that run on ridges across the centre of the county.

Surrey is also important, with the North Downs providing numerous possibilities just south of London. But this is a densely populated area and the most heavily wooded county in England. Also, many of the chalk slopes are steep and high. Yet Denbies, near Dorking, has played an important role and as I drive along the M25 there are numerous south-facing slopes that look attractive to me.

KENT

Chapel Down

Small Hythe, Tenterden, Kent TN30 7NG
www.chapeldown.com Telephone 01580 7766111
First planted 1977 (as Tenterden Vineyard); vineyard area 319 hectares (789 acres)
At the winery: guided tours and tutored tastings, booking in advance recommended, The Swan restaurant; Curious Brewery in Ashford and Gin Works at Kings Cross, London
Sales: cellar door, online, local and national stockists, export
Oz recommends Kit's Coty Blanc de Blancs✶ and Coeur de Cuvée✶, Chapel Down Flint Dry, Bacchus and Chardonnay, Kit's Coty Chardonnay

Frazer Thompson, the boss at Chapel Down, has always looked to the future. I've got a report here that he wrote in 2002, just after he arrived at Tenterden to take over the reins of what was then called 'New Wave Wines'. He was predicting what the next decade would bring. 'English-driven varieties such as Bacchus will be as well known as Sauvignon Blanc is today ... If New Zealand can do it, why can't we?' Well, it's happening. People are now talking of British Bacchus as being our home-grown equivalent of New Zealand Sauvignon and drinking it with brazen enthusiasm. He continues 'Some brands of English fizz will be far more exclusive, original and sought after than their French counterparts.' It's happening. The top English sparklers now routinely beat French Champagnes in competition, and several are already regarded as more exclusive than Champagne. 'English vineyards will become not only common across the south-east, but important centres of employment, hospitality and tourism, as well as a source of national pride.' It's happening. Add to that the south, the south-west, Wales, East Anglia and points north and he encapsulates everything that English and Welsh wine has become.

So if he saw the future clearly in 2002, he sees it equally clearly in 2019. I'm standing in the middle of Kit's Coty vineyard. The soil is pale under my feet. It's late morning in July. The sun is positively hot, but a cool breeze plays among the Chardonnay and Pinot Noir vines that stretch out to the south-west. This is a wonderful site. 40 hectares (100 acres) of what seems to be a jewel nestled under the rough white chalk brow of the North Downs. I've tasted the Chardonnay. It's simply fuller, riper, more exciting than any other Chardonnay from England. 'Is this England's best site?' I ask Richard Lewis, the vineyard manager. He gives me a wonderfully old-fashioned look. We clamber into the 4 x 4 and he drives just along the slope eastward. The aspect to the sun changes by a degree or two. 'This is Street Farm. 40 hectares (100 acres). Planted a couple of years ago. Could be even better than Kit's Coty.' Back in the 4 x 4. Another short drive. Another pale chalk slope, another degree or two change in the aspect to the sun. 'Court Lodge. 44 hectares (108 acres). Could be even better.'

And he's not done. Back out, just a little to the west, over toward Boarley, through a gap in the hedge and another amazing sight.

50 hectares (123 acres) of some of the palest soil I've seen in Britain, perfectly angled toward the sun. Row upon row of new vines. 50 hectares (123 acres). All planted on April 30 2019. 'We've another 60 hectares (148 acres) to plant,' Richard says, waving toward fields thick with red poppies, ready to clear. 'We'll have 160 hectares (395 acres) altogether.' At Boarley. All possibly better than Kit's Coty. I drift back dazed to the sleepy little village of Boxley and contemplate a reviving pint of local ale in the pub. Have the contented residents of Boxley the slightest idea that they may be at the centre of one of the world's great new vineyard areas on the chalky slopes beneath the Pilgrims' Way? And that all these vines, all these grapes are owned by Chapel Down?

How can one company have so much confidence in their own future and the future of English wine? Particularly one like Chapel Down. Well, it wasn't always so. What is now called Chapel Down started out as a contract winemaking operation in Sussex, that expanded into Kent, retitled itself 'English Wine Group', got into trouble, then hired whizzkid Frazer Thompson to get it back on an even keel. He showed his mettle on his first day in charge. The hire company was trying to repossess the photocopier. Frazer persuaded them that without a photocopier he wouldn't be able to send any invoices to get paid so that he could pay them.

Right from the start he set out to make people take notice. He invented the 'Curious Grape' logo to use on the labels of the unfamiliar grapes like Schönburger and Ortega that the company had quite a lot of. He used to send out mad press releases about 'Curious' involving rappers and golfers and the Swiss sailing team. Once he went on about the American spy chief Edgar Hoover and a floral twinset. He even brought the Queen into things. 'Why?' I asked. 'Anything to get people talking about English wine instead of walking by on the other side of the road, clutching a glass of indifferent Chablis or Bordeaux. We just had to get noticed.'

The 'Curious' logo is now applied to the brewery, and the wine side of the business is called Chapel Down. From that unpromising start in 2001, Chapel Down now makes more wine than any other British operation, routinely aiming for a million bottles a year, and in 2018 making well over two million bottles. It's a publicly listed company, and has also championed crowd-funding as well as traditional fund-raising to fund expansion – and, of course, to create legions of loyal Chapel Down drinkers out of thousands of small shareholders.

Because they can use economies of scale, like Denbies, they've provided a lot of wine to supermarkets, and they used to offer sparkling wine far cheaper than anybody else. They still do supermarket business, but they don't do it cheaper than everybody else. Their basic white blend – Flint Dry – is actually a shining example of how to use your less sexy juice to make a delicious, scented, bargain-priced white that just oozes Englishness. It's widely available, and it's not expensive. Their Bacchus is reliably one of the freshest in England, the uplifting scent of spring flowers and elderflower with a squeeze of lemon zest – does that sound alluring? Chapel Down do it, for not too much money. And I haven't forgotten. Kit's Coty Chardonnay is a game changer for England, and not cheap. Kit's Coty Coeur de Cuvée fizz is intense and memorable – and £100 a bottle. People buy it. There are 40 hectares (100 acres) of Kit's Coty. There are 244 hectares (603 acres) from that magic slope yet to bear fruit. That's an awful lot of wine to come and the wines will be good. The challenge will be to get people to buy them. Frazer's been pretty good at that so far.

Gusbourne

Kenardington Road, Appledore, Kent TN26 2BE
www.gusbourne.com Telephone 01233 758666
First planted 2004; vineyard area 90 hectares (220 acres)
At the winery: vineyard and estate tours, The Nest tasting room, cheese and charcuterie plates, other food available for prebooked tours
Sales: cellar door, mail order, online, local and national stockists, export
Oz recommends Blanc de Blancs∗, Rosé∗, Guinevere Chardonnay and Pinot Noir

---

I heard the rumours. There was a new, exciting, ambitious high end producer in Kent. About time, I thought. Kent has always seemed to me to be surprisingly slow in building on the hard work of Chapel

Down in particular. And Kent seemed to have so much fantastic chalk soil which, every year, I was convinced must start sprouting vines, and it never did. But this was 2010 and it sounded as though that chalk was finally starting to be put to the plough in order to reveal its glories.

Except it wasn't. I was tasting the new releases from an outfit called Gusbourne, above all their amazingly rich but dry, imposing Blanc de Blancs, with the thrilling flavours of sticky crème fraîche and croissant crust that you often hear tell of in Champagne, but rarely find. Gusbourne. I didn't know it. Which chalky suntrap did this come from? The North Downs is full of likely sites, from Dover all the way to Gravesend and across to Sevenoaks. And then I looked closer. Appledore. Appledore? But that's down on the marshes, isn't it? That's a piece of Kent where few people live, few people visit, where the mists linger late into morning and the sodden soil squirts and squelches underneath your boots. Flat, strangely forbidding, dominated by the smugglers' paradise of Romney Marsh, only famous for sheep and will o' the wisps and shadowy figures going about their silent business.

Even as a Kent boy, I just about never visited this strange, unalluring corner of the county. But then I saw a photograph, taken from Gusbourne's supposedly finest vineyard, Boot Hill. I had been here before. I'd wandered over from the metropolis of Tenterden with my friend Mike one January day. I think we'd been shooting, and now we wanted a brisk walk and a foaming pint of beer.

And I do remember standing on a low rise, a very low rise of land near Appledore, just above the Old Military Canal. The sun hung low in a cold blue sky, that mid-winter sun which is so bright it near blinds you with its glowing whiteness but is suggestive of the brittle tingle of an ice world untouched by warmth. Steamy breath, a feathering of snow glistening in the sharp raw light, eyes searching for any sign of a pub across the crisp silent distance stretching away to the sleepy slovenly surf of Dungeness beach. Long before any vines, I had stood and pondered, right there on Boot Hill. Right where Gusbourne's best vines grow. And there was no chalk then, and there's no chalk now.

But there's something equally important, maybe more important. There is a vision of greatness and a determination to produce England's best wine, even if the soil is not the nation's best. And a belief that the best way to do this to combine a painstaking old-fashioned view of raising vines and methodically turning their fruit first into wine, then into sparkling wine. But doing this with the most modern equipment. And above all, never losing patience, never hurrying, never losing sight of the vision of flavour that spurs them on. That crème fraîche richness, that roasted hazelnut, that croissant crust …

All of this from a 'turnip patch'. That's not my description, but it's what Andrew Weeber, the South African founder of Gusbourne, called this parcel of sullen clay that he bought in 2003. He'd been part-owner of a vineyard in South Africa and obviously got a bit homesick for it over here in the UK. But his daughter lived in Appledore and every time he visited he was struck by how warm the weather was, how the sun always shone.

Ah, now it begins to make sense. In South Africa they don't have many of the classic European soils to grow grapes on: climate, weather conditions and things like water availability are of massive importance there. And Appledore isn't quite sucked into the Marsh. There is a tiny escarpment called the Saxon Shore – the old shoreline – and Weeber found a property that straddled this shoreline and rose from virtually sea level down by the old canal up to about 45 metres (148 feet) at the top of Boot Hill.

The soil? Tough, hard, compacted sandy clay. But he swore that when this clay cracks in summer, it breaks off all the vine's surface roots and forces them downward. Which might explain why Gusbourne's wines attained such depth of flavour right from the get-go. It's not the classic way. But the slope is gentle, the aspect to the summer sun ideal and the gentle breeze from the Channel is warmer than the north Kent breezes off the Thames and the dark North Sea. So they planted Chardonnay, Pinot Noir and Pinot Meunier. Some were Champagne clones only good for making sparkling wine, but the best patches of land were planted with clones from Burgundy, good for fizz, but also able to linger on the vine and make really tasty still wines, when the sun shines. And it does. This is Kent. Barely six miles away is the Dungeness beach – and that's the only certified desert in Britain.

Charlie Holland is the winemaker in charge of this project and he's got the vision. Like all English winemakers he hankered after some chalk, and Gusbourne then bought two chalk sites near Goodwood in Sussex to satisfy that craving. But right from the start that mild, unprepossessing Boot Hill slope has not only provided great fruit for fizz but has also been the heart of the impressive still Chardonnay and the Pinot Noir.

It's the Pinot Noir which turns heads and in years like 2014 and, to a lesser extent, 2018, this cautiously full-coloured red can have a suggestion of violets in its scent, a fruit whose cranberry and rosehip leanness just hints at softer strawberry sweetness for those who wait, and whose smoky, saline savouriness really might twirl and flirt and bit by bit entwine itself with the Boot Hill fruit to finally make what some might call England's first Burgundy. But I'd rather hope it's simply England's first great Pinot Noir.

Hush Heath Estate/Balfour Wines

Five Oak Lane, Staplehurst, Kent TN12 0HT
www.hushheath.com Telephone 01622 832794
First planted 2002; vineyard area 53 hectares (130 acres);
total average annual production: approx. 425,000 bottles
At the winery: self-guided and guided tours, 200-person tasting room,
large decking and balcony looking over the estate, regular events
Accommodation at Hush Heath's own local pubs, The Goudhurst
Inn, Tickled Trout and Ship Inn Rye
Sales: cellar door, including Jake's beers and ciders, online,
local and national stockists, export
Oz recommends Balfour Skye's Blanc de Blancs✻, Balfour Brut
Rosé✻ and Balfour Liberty's Bacchus

I remember the judges' amazement in the 2008 International Wine Challenge when we were faced with this pale, salmon-coloured sparkling wine, incredibly dry, understated, sophisticated, surely expensive – and it was English. Hush Heath. It was quite the most elegant sparkling pink in the competition, and from a new Kent estate that none of us had ever heard of.

Richard Balfour-Lynn is not a man short on self-confidence, but I bet even he was pretty close to being struck dumb when he got word that his 2004 rosé, his first vintage from grapes only planted in 2002, had won a gold medal in this global competition as well as the Trophy for Best English Wine. The 2-hectare (5-acre) vineyard he had planted on his country estate was supposed to be a hobby. And now it was national champion. Well, in a way, it couldn't have happened at a better time. Richard was a property tycoon but in 2008 the property market crashed by the biggest percentage fall on record. Richard escaped with his skin intact, but his shirt pretty torn. Could that trophy for his rosé be a sign that he needed to slow down and take a new direction?

Which is what he did. He was an hotelier as well as a property guy, and this side now came to the fore. He decided he would turn Hush Heath into a prime winery-cum-tourist attraction. And he would be at the centre of it. Meeting Richard in London, it's easy to feel he hasn't quite got the 'big businessman' bug out of his system. You have to go to his estate. But he would entirely agree. He'll meet you at Marden station in deepest Kent, drive you off through country lanes that speak of Kent's plump, prosperous Weald, and then, without noticing it, you're on the estate, of which not that much is vineyard. The estate is 162 hectares (400 acres) in total, and the most significant feature is the dense, dark, mystical 500-year-old oak forest which reluctantly allows a modicum of vineyard and orchard to intrude on its inscrutable serenity.

And until 2018 not much of it was winery, either. After the initial success of the rosé fizz in 2008, we waited expectantly for this new star to shine more brightly, but it didn't, and although he built a small winery, employed a very good winemaker and began to work on tourism, the whole venture seemed to be marking time. I suspect Richard felt that, too.

When Richard started out he said 'we'll never make more than 10,000 bottles'. That plan wasn't working, so in 2015 he came up with a new one which looks like being a successful hybrid of his previous business flair and his new-found satisfaction in his Kent estate.

Out went all idea of 10,000 bottles. The new plan was to aim for 750,000 bottles. No half measures here. There were only about a further 12 hectares (30 acres) really suitable for vineyards on the estate, so he'd plant those, and go into partnership with proven local fruit farmers to plant at least another 40 hectares (100 acres). There was a quality factor, too. The Hush Heath soil is heavy clay, which they have done as well with as they can. The new farming partners all had south-facing slopes of warm greensand soils. Their grapes could add an entirely new dimension to the range of wines which was now spreading successfully away from sparkling wines to still wines. Typically the still Nannette's rosé is already one of the best wines – the Estate fruit is clearly suited to making pinks. The cricketing legend-turned wine entrepreneur, Sir Ian Botham, thinks so too – he's brought out a still and sparkling pink from Hush Heath fruit.

Part of the new plan was that Hush Heath would really take wine tourism seriously. Many English wine estates now know how important a part of their business tourism is. Hush Heath sets out to make it an irresistible part of their activities – they call it 'adult tourism'. Along with an impressive expansion of the winery in 2018, there's a new visitor centre which is airy, spacious, relaxing, expensively finished in wood, copper and stone; it has both a wide balcony and an expansive tasting and relaxation area with a sweeping vista through the vines up to the Tudor manor house, and, always, the deep, dark forest.

It could be Napa Valley rather than England – and I'm sure that's what Richard had in mind. The difference is that wineries in the Napa are often designed simply to sell. Hush Heath is designed to invite you in, and, as long as you respect what Richard says is, after all, our home – the vineyards, the wild flowers, the carpets of bluebells and white anemones in the spring and wild forests which have been left to brood since Henry VIII was on the throne, are as much a part of the experience as the wine.

Simpsons Wine Estate

The Barns, Church Lane, Barham, Canterbury, Kent CT4 6PB
www.simpsonswine.com Telephone 01227 832200
First planted 2014; vineyard area 30 hectares (74 acres);
total average annual production: approx. 250,000 bottles
At the winery: vineyard and winery tours, Glass House tasting room,
supper club wine dinners, Fruit Chute helter skelter, events
Sales: cellar door, mail order, online, local and national stockists,
export
Oz recommends Chalklands Classic Cuvée✳*, Roman Road*
Chardonnay, Gravel Castle Chardonnay, Railway Hill Rosé and
Derringstone Pinot Meunier Blanc de Noirs

'I came to Kent after 17 years in France. I had no family ties. I
came because I believe that in Kent we can have, not the best
vineyard sites in England, but the best vineyard sites in the world.'
This was Charles Simpson speaking at the 2019 Canterbury Wine
Show, and the packed hall literally thrummed with excitement.
Kent had seemed to be the bridesmaid of English wine for too
long. Sussex and Hampshire had stolen all the plaudits, and even
Surrey, Dorset and the Thames Valley had forged ahead, leaving
Kent in their wake. But wasn't Kent the garden of England? Wasn't
Kent fruit more famous than that of any other county? And didn't
the fabled seam of chalk that ran through the greatest vineyards
in Champagne also rear its head at Dover and then march into
the Kent countryside as the Kent Downs and the North Downs,
scattering perfect, south-facing slopes as they went? And didn't the
sun shine more brightly in Kent as well?

I would agree with all those statements. But then I am a Kent boy.
And I shared the excitement at what Charles and Ruth Simpson
were doing when they planted two broad slopes of the Elham
Valley, just south of Canterbury at the village of Barham. I had
lived in the next village. I knew exactly the fields they had planted
because I had tramped through them as a child after church on
Sundays, working up an appetite for lunch. And I became engaged
just down the lane from what is now the Simpsons winery. The
bond didn't last long. About 10 minutes, but I was only four years

old. I know, ruthless heartbreaker. I'm always a bit nervous in Barham if I see someone lurking in the shadows.

But what brought the Simpsons to Barham? Well, there is one very important point. They could well be the most professionally qualified of any of the new wave of pioneers who have established vineyards and wineries in Britain. They've done it all before. They bought a property in southern France – Domaine Sainte Rose – in 2002. It's tough to make a success of high-end wine in France's Languedoc region, but they've done it. As they said, 'we've made all our mistakes already' and, by focusing on quality and on export – they didn't sell any of the Sainte Rose in France – it's now a thriving and profitable operation.

But family affairs were drawing them back to England, and when they began to drink English sparkling wine and be amazed by its quality, and then to realise how little had yet been achieved over here, and how great the potential was both for profit, but also for creating a family business with a real legacy to it, it became a case of finding the right piece of land. And they found it. One slope facing south-west just below the old Roman road from Dover, surrounded by protective woods and with the chalk barely 30 centimetres (12 inches) below the surface; the other slope, just over the old railway line, this time facing south-east but with the same soils, the same breezes running down the valley, the same Kent sunshine to start the budding and the flowering ahead of Sussex, and to get the grapes to a ripeness where you can make still wines as well as sparklers – before Sussex.

Running a wine estate in France had given the Simpsons excellent contacts and so they were able to clear and plant their vineyards, starting in 2014, considerably more cheaply than most other English outfits can. And in a year like 2018, when their crop went from 25 tonnes the year before to 200 tonnes in 2018, they could make the telephone call to France and get extra tanks when others couldn't. And their French wine experience gave them very particular views about what the strengths of their English wines should be. 'After 17 years of making wine in southern France, there's one thing we really know – texture and mouthfeel.' Although Charles swore that he would never make still wine, he couldn't resist having a go in the bountiful 2018 harvest and came up with beautiful Chardonnay, good Pinot Noir and a fascinating still pink Pinot Meunier – almost as a sideline as the first sparkling releases became ready for market.

They know how to sell their wines, too. Marketing can seem like a foreign language to many of the English wine pioneers. But the Simpsons are already successful in the USA, Scandinavia, China, Hong Kong and Japan – and swear that markets like these do get the message that English stands for quality. When I was last down in Kent, Charles had just been at a meeting of all the UK ambassadors at the Foreign & Commonwealth Office in Whitehall. They were serving five British gins and three English sparklers – including Simpsons – and Charles said the one question he simply had to ask was why are any of you still serving French Champagne and not English sparkling wine? An ever-increasing number of them aren't.

SUSSEX

Bolney Wine Estate

Foxhole Lane, Bolney, West Sussex RH17 5NB
www.bolneywineestate.com Telephone 01444 881575
First planted 1972; vineyard area 16.2 hectares (40 acres);
total average annual production: approx. 250,000 bottles
At the winery: tours, tasting room, restaurant
Sales: cellar door, online, local and national stockists, export
Oz recommends Classic Cuvée, Bolney Bubbly*, Blanc de Noirs*,*
Foxhole Bacchus and Pinot Gris

It was front page news. Jonathan Maltus, a top producer in Bordeaux's St-Emilion region, was returning to his home county of Kent to plant St-Émilion's Merlot vines. This was completely plausible and I think I believed it. Then I noticed the date. April 1. A very good April Fool's Day joke. So no one was really going to plant Merlot.

Maybe not – yet. But Sam Linter had already planted Merlot years ago, at her Bolney Estate north of Brighton. 'Sussex is the sunniest county,' she said, only ever so slightly sniffily. 'And Bolney is the earliest ripening site in Sussex,' she added. So, any chance of a taste of her Merlot? Not a chance. She'd planted her Merlot on a warm, south-facing slope right by the winery. Year after year she struggled to get even a decent rosé out of the grapes. She did once make a red. Could I …? The grimace on her face told me I really wouldn't want to. She ripped out all the Merlot and planted Chardonnay instead.

I wouldn't put it past her to try again. She loves red wine. She reckons Bolney makes more red wine than any other winery in England. But just at this juncture in the solemn march of climate change, she's focused on Pinot Noir. Her parents planted some Pinot Noir as long ago as 1972 when they bought what was then a pig farm along with some chicken sheds. Sam says 'We've been running the gauntlet with Pinot Noir for 47 years,' and her Pinot Noir reds do seem to epitomise a pale, cool English style. She says the key to English Pinot Noir is not to try to make it taste like anything else. You must embrace the lighter, fresher style. If you try to work the skins too much for greater concentration and colour you won't be able to avoid bitterness. 2018 was a vintage where she could have been tempted to give the whole Pinot thing a bit more wellie, but she didn't. The wine is pale, ethereal, scented with rosehip and redcurrant, pepper and raw red plums – nothing to do with Burgundy, fragrantly English.

And Bolney could well be the largest producer of Pinot Noir in Britain. In 2018 Sam made 22,000 bottles of it, but her passion for red doesn't stop with still wine. She uses the Dornfelder grape to make a very refreshing, lean but tasty sparkling red, surprisingly rather than shockingly dry, and pretty spot on as a chilled thirst-quencher to go with something like lamb chops.

To be honest, the main Bolney vineyards are pretty adaptable – they have 16.2 hectares (40 acres) near the winery, and, since 2018, a further 27 hectares (66 acres) just a mile or two away at Pookchurch (love the name). And so far they seem to be able to ripen pretty much everything – except Merlot – that has been planted on them; there's a lot of warm sandstone mixed in with the soil. These new hectares put Bolney into the 'bigger than boutique' class of winery. Sam's objective is to get to an annual production of 300,000 bottles – a few more abundant years like 2018 and Pookchurch coming on full stream will make 300,000 bottles the norm. Yet Sam says she doesn't want to take on the Big Boys. She wants to keep Bolney boutiquey, to keep it in the family.

But what do you do? She's caught in the old bind – to stay the same, you have to change. She's already built a brand new winery area for the 2019 vintage. And she says at 300,000 bottles, unless she magically doubles her sales, she's going to need a new warehouse. Yet this prospect seems to please her. She'd love to give her wines more time to age before being sold – she'd like to give her fizz 12–18 months more than she currently does to soften up and develop its flavours. And that will cost money.

Bolney hosts about 10,000 visitors a year, but with attractive tasting and drinking areas looking out onto the vines, and a delightful sylvan setting down leafy lanes north of Brighton, that could easily increase. There's already a Foxhole Gin made from wine pressings, and a vermouth using estate produce like sloes and elderberries. And putting your wine in front of the right type of people helps. You might have seen Bolney listed in British Airways First Class. And in 2015 and '16 you could have been drinking Bolney Pinot Gris at the Wimbledon Lawn Tennis Championships in London. Because in 2018, boutiquey, family-run Bolney became the first English winery to gain distribution in mainland China.

Breaky Bottom Vineyard

Whiteway Lane, Rodmell, Lewes, East Sussex BN7 3EX
www.breakybottom.co.uk Telephone 01273 476427
First planted 1974; vineyard area 2.4 hectares (6 acres);
total average annual production: approx. 11,700 bottles
At the winery: visits by appt, minimum of 6 people, tastings
Sales: cellar door, online, local and national stockists, export
Oz recommends Cuvée Gerard Hoffnung✳, Cuvée Oliver Minkley✳
and Cuvée Peter Christiansen✳

I was leafing through the results of the WineGB Awards in 2019 and one name came up again, and again. And again and again. And I felt a tinge of emotion. The winery was Breaky Bottom and the owner was Peter Hall. And I found myself almost crying, oh, Peter. You're still here. How are you still here? And you're still crafting unbelievably original wines and what's more, you've actually got enough bottles to enter into competitions. And I bet you haven't changed a jot. You probably haven't even changed your proudly seedy French beret and matelot's jersey, either. And I bet you are still pulling for all you're worth on those odorous French Gauloises – or do you roll your own Old English shag? And they just behave and smell like French Gauloises? Yes, perhaps you do.

Breaky Bottom, which supposedly means a bracken-infested tough downland valley, ought to represent everything that is tranquil and perfect about a family-run English vineyard. But it's not called Breaky Bottom for nothing. The South Downs look like a paradise as you gaze toward them from the sylvan safety of prosperous mid-Sussex. So why are there no roads criss-crossing them? Why is there barely a village, scarcely a hamlet? Because the Downs are savage and raw when you clamber to their tops, rearing hundreds of feet up above the cosy towns of Brighton, Seaford and Eastbourne. Winds howl, frosts fall, sunshine shrivels in the face of salty sea frets and wildlife takes unkindly to human intrusion and plots vengeance day and night.

And yet, and yet … there must be havens of tranquility in the clefts of these bleak uplands. Well, yes, there's one at very least. It is almost impossible to get to because of the antediluvian state of the access track and almost impossible to leave because magic reigns in this tiny spot. And magic is both light and dark. About 10 years ago I spent the summer ranging through Britain in a caravan with my petrolhead friend James May. With his formidable off-road driving skills – largely honed by the number of times he's spun off the track in his Top Gear days – we did find Breaky Bottom, and we did manage to leave. I'm not sure he completely shared my enthusiasm but I had known this place and its wine for over 20 years by then. So I sat and wrote a few notes that went something like this:

'There is no more beautiful vineyard in Britain than this hidden eyrie. A tiny golden spot of fairy magic in a fold of the Sussex South Downs. The world has grander sites, more opulent spreads, more dramatic escarpments trailed with vines, but Breaky Bottom is about heart and soul. So delicate, so fragile, so precious and held in the loving, caressing bulldog hands of Peter Hall. You take a glorious but bone-shaking drive across the crest of the Downs, with Beachy Head jutting seawards far off to your left, and suddenly, way below you, in a tiny valley so tight it's more of a cleft than a fold in the hills, you see two patches of vines – one small, one even smaller, dappled by Sussex sun. And you see a flint-faced cottage so tranquil that hard-eyed bankers would weep to possess it.'

They never will. Peter Hall, whose nut-brown gleaming face radiates calm and generosity, has fought for these priceless acres – barely 2.4 hectares (6 acres) of them – since 1974; he had stumbled on them while working as a shepherd. He has been felled more than once by lack of money – in some years he has made no wine at all, at other times his wine or his crop has been destroyed – and bureaucracy, nature, and his own slightly contrary outsider's disposition have repeatedly knocked him to the floor. When I first visited Breaky Bottom he was broke. Actually broke and working for cash in a nearby Lewes restaurant to pay his grocery and tobacco bills. He was once flooded out five years in a row. Every kind of wild animal or bird or fungus has had a go at decimating his crop. But he is still here.

Overleaf: Ah, my old favourite – Breaky Bottom's tiny, idyllic vineyard, nestles in a protective fold of the South Downs near Lewes.

And he still welcomes his visitors and opens his wines and drinks with you – I don't think he has a spittoon. So tarry a while and you will learn about wines, about vineyards, about life – yours, his, what life has been, what life could be ... Peter hates to leave this place he loves. And so will you. Don't look back until you're well up the rutted track toward the windswept heathland of the Downs. Peter says that when he found Breaky Bottom, he was entranced because it possessed 'a kind of magical sufficiency'. It still does.

Breaky Bottom initially gained its fame for clean, bone-dry whites from Seyval Blanc and Müller-Thurgau in the 1980s when few English wines were clean and few English wines were dry. His Müllers were peppery, leafy, crackling with grapefruit tang – and delicious. The Seyvals uncovered flavours of pink grapefruit, peach skins, fresh limes and even floral scent that no one else had unearthed. Other English winemakers called Seyval 'hideous'. Only Peter Hall calls it 'elegant, sweet-natured'.

Peter doesn't make still wine any more. All the Müller-Thurgau vines have gone, and some of the Seyval, being replaced by Chardonnay, Pinot Noir and Pinot Meunier. Peter was always keen on fizz – his mother was from the Champagne Mercier family – so now that's all he makes, various cuvées under a variety of unlikely names. One of his 2019 winners was Cuvée Gerard Hoffnung, named after a brilliant comic musician whom Peter knew. Even the wines made solely from Champagne varieties taste different at Breaky Bottom. But Peter also soldiers on with Seyval Blanc in some of the blends, and they really are like no other English fizz, with their flinty, pithy kick. Shouldn't he rip up the Seyval? Why? It's been a loyal friend for 40 years, and that matters to Peter Hall.

Nyetimber Vineyard

Gay Street, West Chiltington, West Sussex RH20 2HH
www.nyetimber.com Telephone 01798 813989
London office: 020 7734 8490
First planted 1988; vineyard area 280 hectares (692 acres);
total average annual production: approx. 400,000 bottles
At the winery: ticketed open days (including tasting)
Sales: online, local and national stockists, export
Oz recommends Blanc de Blancs✳, Rosé Multi-Vintage✳ and
Tillington Single Vineyard✳

'You should grow apples. There's no future in vines.' If the Ministry of Agriculture's vine expert who came down to Nyetimber in deepest Sussex in 1987 is reading this, I hope his ears are burning. An American couple, Stuart and Sandy Moss from Chicago, had bought a beautiful but rundown estate called Nyetimber in 1986. The house and gardens were going to need a lot of love, but the adjoining land was on a relatively warm soil type called greensand, much prized in Champagne for the quality of the grapes it produces, and the slopes faced south. Good for grapes? Call in the man from the Ministry for advice. He'll help. He didn't. 'We didn't move 4000 miles to grow apples,' growled Stuart.

But perhaps he did help. The Mosses delighted in being cussed. Tell them they can't do something, they'll be even more determined to do it. We can't grow grapes and make wine? Of course we will. And there's another thing. Over the last 30–40 years, as climate change, at first slowly, but now rapidly, is transforming many areas of the globe that were dismissed out of hand as being entirely unsuitable for vines, one refrain comes up again and again. 'It's too cold for vines. Plant apples.' That's what the experts said in Australia's Tasmania. That's what they said in Elgin in South Africa. That's what they said in Canada and in New York State. That's what they said about the whole of South Island of New Zealand. Nowadays every time I hear an agricultural expert say that a place is only good for apples, I'm tempted to ring up a vine nursery and order 100,000 vines to plant, because the results will clearly be superb.

this title before. She says that she feels confident about being able to handle a harvest volume of a million bottles without compromising quality. She puts a lot of her wine aside to age as a 'reserve' – usually 25–30 per cent, and more in a big vintage like 2018. This allows her to make the most popular Nyetimber wine – Classic Cuvée – as a multi-vintage blend that contains components of vintages going back 10 years. In terms of achieving consistency and quality, this is wonderfully effective and keeps Nyetimber effortlessly at the top of the English quality ladder. It must be expensive to do this. Yes, it is. Luckily Eric Heerema is up for the challenge.

Plumpton College

Ditchling Road, Lewes, West Sussex BN7 3AE
www.plumpton.ac.uk Telephone 01273 890454
First planted 1988; vineyard area 9 hectares (20 acres);
total average annual production: approx. 18,000 bottles
At the winery: visits by appt
Sales: online, local and national stockists
Oz recommends Brut Classic✴, Bacchus and Rosé

I'm surrounded by a gaggle of inquisitive 13-year-olds on my first visit to the wine department of Plumpton College, England's increasingly famous wine school. They're all here for the 'Big Bang' – an afternoon of donning white coats and lab spectacles, pressing grapes, messing around with refractometers and the like and probably ending up exhilarated and mucky – just what kids love. But did they know anything about wine? Only one said 'yes; put it in front of my parents and it's gone.' Well, that's a start.

But who knows how many of these schoolchildren might catch the wine bug from visits like this? Plumpton College is Britain's only vineyard and winemaking school, and it's the only establishment in Europe that teaches the wine business and the production of wine in English. Chris Foss, the boss from its foundation in 1988 until 2019, reckons there are now few wineries in Britain that haven't benefitted from Plumpton graduates, and I've met them on my travels in South Africa, Australia and New Zealand as well as in France and Spain.

The wine course didn't start very auspiciously. Chris Foss had been making wine in Bordeaux in France – in some years up to 500,000 bottles – but had come back to England in 1984 to look for a vineyard manager's job. These weren't thick on the ground in 1984, and the Plumpton Agricultural College was just wondering whether to dip its toe into the wine and vineyard business. The college wasn't exactly going to shower gold on the idea but they had two rows of vines, and Chris seemed content to make do with a desk and an old chicken shed and some leftover glass demijohns.

But Chris was undeterred – or just desperate to get away from the smell of poultry. The two rows of vines became 0.4 hectares (1 acre) and is the now the heart of the experimental college vineyard, which aims to include every trellising and pruning system known to man, as well as every grape variety and rootstock that might reasonably be expected to have a future in Britain. There are also 10 hectares (25 acres) nearby which Plumpton operates as a commercial vineyard, bringing in a very useful £150,000 a year to fund the College programmes.

And the wine department has had a purpose-built Wine Centre since 2006, which is improved on every time a new sponsor is prepared to put up the cash. The winery itself is small but smart, with tanks as tiny as 200 litres and as big as 5000 litres, so that the students can make as many experimental cuvées as possible – which they will then sell, unless they are filthy, in which case they get thrown away or, knowing students, another use for them is 'discovered'. There are wooden barrels – they were barrel-fermenting Ortega and aging some red Pinot Noir last time I was there. There is also a row of *qvevri* earthenware jars half buried in the soil outside one lab to try to make orange wine. And there is a miniature sparkling wine unit so that the students get the hang of the very specific technical skills needed for sparkling wine.

Plumpton now offers a less academic Diploma Course because there is a real demand for qualified vineyard and winery staff, but the main activity is at degree level. Since 2018 it has worked in partnership with the Royal Agricultural University in Cirencester, Gloucestershire. There are 30 students studying winemaking, 30 doing wine business studies and 20 on a master's course.

Research is increasingly important and will be invaluable for the future of British vineyards, particularly as regards the effects of climate change and how to combat disease in British conditions. On one acre of the vineyard they have planted 40 different recently created vine varieties that are resistant to downy and powdery mildew. In a damp maritime climate and with a global move against the use of fungicides we might well be seeing more of these in the future, and the groundwork will have been done at Plumpton.

This is all very high-powered. Yet as I wander through the rooms and corridors, the place seems reassuringly down to earth. Every wall is plastered with posters – they're all to do with important subjects like 'yeast metabolites as lactic acid bacteria', but they lift my spirits rather than oppress me. I see a line of Bodum cafetières – 'ah, we use them to improvise submerged cap fermentation experiments'. There are vines growing in portable wheelie bins. There's a dubious-looking bottle on a bench, hand-labelled 'Cider Cloudy'. And the demijohns aren't being retired just yet. There's a bunch of them bubbling away in a corner – that's a Muscat fermentation experiment. They bought the grapes from the local greengrocer.

Ridgeview

Fragbarrow Lane, Ditchling Common, East Sussex BN6 8TP
www.ridgeview.co.uk Telephone 01444 242040
First planted 1995; vineyard area 6.5 hectares (16 acres); grapes sourced from a total of 85 hectares (210 acres); total average annual production: approx. 400,000 bottles
At the winery: open daily for tours and tastings, prebooked tours via website, for group bookings contact winery direct, Wine Garden open May–Sept with picnic hampers available, pop-up chef dinners, Family Easter trail, annual Ridgefest festival in August
Sales: cellar door, online, local and national stockists, export
Oz recommends Blanc de Noirs*, Blanc de Blancs*
and Rosé de Noirs*

Mike Roberts loved a good view. He had built a serious computer company from his suburban front room, ending up with 500 employees; and he had always taken his best salesmen to Champagne as a reward. They would stay at the Royal Champagne Hotel and, with a glass of fizz in their hands, would gaze out over the vines toward Épernay, the dozy River Marne flecked with mist and lined with poplars, and across to the ancient Abbey of Hautvillers, where Dom Pérignon was supposed to have invented sparkling wine. It's possibly the most beautiful view in all Champagne.

Mike sold his computer company, breaking free from an increasingly corporate business environment, and gradually his love of the Champagne region nagged away at him as he considered what to do next. Could he try to make Champagne-style sparkling wines in England? He was friendly with the Tenterden (now Chapel Down) people, and they thought you could, and so when a farming estate came up for sale just a couple of miles down the road from where he lived ... could he do it? What was that rumour he'd heard about some mad American couple doing it only a few miles to the west of him in Sussex, at what was to become famous as Nyetimber? So he drove down Fragbarrow Lane, just outside Ditchling Common, and there it was – the view. Plant these fields with vines, and after a hard day's work, just pause here, as the sun sets in the west, his own vines stretching away toward the chalk slopes of the South Downs, the soft breeze dying as the ripening crop licks one last ray of warmth from the day. Yes, he could do it all right. He became convinced that within a generation England could be making the world's greatest sparkling wines, and he wanted to be part of the journey.

It was 1995 when the Roberts family planted their field and named their winery, Ridgeview Estate. Mike always prided himself on being a risk taker, but he always had a plan. And at Ridgeview it was simple. Only plant the same grape varieties as they grow in Champagne – Chardonnay, Pinot Noir and Pinot Meunier. Right from the start, build a fit-for-purpose winery for making the wine, storing it, and turning the still wine into sparkling wine – using the same equipment as in Champagne. And only make sparkling wine, and only use the traditional method as in Champagne. In other words, aim high, don't compromise.

Overleaf: This is the view from Ridgeview's tasting room of Chardonnay vines warmed by the afternoon sun and protected from wind and rain by the South Downs beyond.

That attitude bore fruit very quickly. They made their first Ridgeview wine in 1996 – they had to buy the grapes as their own vines weren't cropping yet – and it went and beat allcomers to win the UKVA's Award for Best English Wine in 2000. Mike Roberts was asked to take it to Australia and present it at the 2000 International Cool Climate Symposium to show what the potential for Champagne-style sparkling wine was in the UK. Mike must have been delighted by his success, but shocked as well. Certainly he had planned to make a high end product – one of his early rules was that Ridgeview should only be listed in local quality restaurants and retailers. But it was a sort of retirement project. He was thinking of, perhaps, making 20 or 30,000 bottles a year – the sort of amount you could hand sell. He did not set out to become the most significant brand in the world of English sparkling wine.

But when you win the big awards and start travelling the world banging the drum for English sparkling wine, the phone starts ringing. Everyone wants a piece of you. And when his son Simon began winemaking alongside Mike in 1998 and daughter Tamara joined him in 2004, the modern Ridgeview started taking shape. Initially they planned to take production up to 100,000 bottles, then 150,000, but it gradually grew to 250,000, and in the massive 2018 vintage, they probably made 500,000 bottles, as well as using their facilities to make up to 100,000 bottles under contract for other wineries and vineyards, and to make own-label fizz of very high quality for such companies as Laithwaite's and Marks & Spencer.

And you're not going to get that much wine out of the original estate vineyard, barely 7 hectares (17 acres) around the winery. So Ridgeview now sources grapes from a total of 85 hectares (210 acres), including partnership vineyards, the bulk in Sussex but some as far afield as Suffolk and Hertfordshire. These will add variety to the flavours of the wines, but also provide security – bad harvest conditions are unlikely in every site.

Yet the most important development for Ridgeview was entering into a partnership with the Tukker family of the Tinwood Estate near Chichester who were making their living as one of the largest suppliers of salad crops to the British supermarkets. That's a high pressure, low margin business to be in. They owned a large chunk of land, flint soil over pure chalk, that would prove ideal for high quality grapes. The Tinwood vineyard was just what Ridgeview needed and just what the Tukkers wanted to create.

There are now more glitzy brands of English sparkling wine than Ridgeview, but there's no more important brand. They understand the importance of tourism. They understand the need to export (I helped launch their 2008 vintage – in Paris!). They already sell a quarter of their output abroad. And they are often the first brand newcomers to English fizz get to try. Tamara says, 'We're the fun, celebratory brand, not the elitist.' Sort of Moët & Chandon rather than Krug. I would agree with that. Britain needs a top quality brand without too many frills, which even so has managed to be served at four state banquets. And they won't be able to stop expanding, either. When I visited in 2019, the place was a dust-clogged building site as new capacity was going up in time for the vintage. But I climbed the stairs to the tasting room, and looked south to the Downs. Mike died in 2014, but don't worry, Mike. Your view is still there.

SURREY

Albury Estate

Silent Pool, Shene Road, Albury, Surrey GU5 9BW
www.alburyvineyard.com Telephone 01483 229159
First planted 2019; vineyard area 5 hectares (12½ acres);
total average annual production: approx. 25,000 bottles
At the vineyard: open weekends, wine tastings, cheese
and cured ham platters, fine dining pop-up events, opera
Sales: cellar door, online
Oz recommends Silent Pool Rosé and Chardonnay (biodynamic)

If I were asked to recommend a really good marketing ploy for an English wine, I think I would say 'being served on the Royal Barge for the celebration of the Queen's Diamond Jubilee in 2012' would come pretty close to top of my list. Especially if it was the first wine I'd ever made, from my first crop of grapes. Well, that's what happened to Albury Estate with their Silent Pool Rosé. Nick Wenman, the owner, was pretty pleased, and lot of rather smart outfits like the Royal Opera House were suddenly on the telephone asking for some too. But nothing gave him greater pleasure than selling 300 bottles to Raymond Blanc, owner of the Michelin-starred Le Manoir aux Quat' Saisons. 'My greatest achievement,' says Nick, 'selling 300 bottles of English wine to a Frenchman.'

At first sight, Nick Wenman looks to have trod a quite common path in the new wave of English wineries. He made money, from IT in his case, and retired. And then he looked around locally to find some land to establish a vineyard – perhaps a sort of 'mid-life crisis? 'Not really,' says Nick, 'I've been obsessed by wine all my life.' When he won the economics prize at school aged 17, he chose Hugh Johnson's *World Atlas of Wine* as his book. Not that the first edition gave much encouragement to a potential English vineyard owner. I seem to remember that Scotch whisky got more coverage than English wine. But he was lucky in where he lived. The Surrey Hills are an intensely beautiful, heavily wooded, heavily protected – and expensive – part of England in and below the North Downs. He looked at seven or eight sites, but one took his fancy – at Silent Pool. He discovered that vines had grown there in the 1640s and John Evelyn had talked about the vineyard in his diary of 1670. He liked the look of its clay on chalk slopes running down southward and hemmed in by forest. His consultant Stephen Skelton liked the look of it. And then Nick announced that he was going to run Albury as an organic vineyard. 'No,' said Stephen, 'no.' Not in our damp, unreliable climate. You need herbicides and fungicides. You won't get a crop. This was not the right thing to say to Nick, whose response to being told 'you can't do it' was to say 'yes, I can. And what's more, I'll go the whole hog. I'll go biodynamic.'

Nick now admits he was fantastically naïve to plunge in like this head first. But he says it was something he just felt was right, and if he didn't do it from the very start, he'd never do it. Being biodynamic certainly adds enormously to the burden of running a successful vineyard. For a start he gets a smaller crop, perhaps 20 per cent smaller. He and his small team need to spend 50 per cent more time in the vineyard. He could clear the weeds by spraying chemicals and this would only take 15 hours per season. But he deftly cultivates between the vines and this takes 12 to 18 days per season. He uses less effective organic sprays and may have to pass through the vineyard 24 times in a season as against the 12 times which will usually suffice for a conventional grape farmer.

But he thinks it's worth it. Above all he says, 'it's the endless concentration that makes all the difference. Observation. Always being on the lookout for the early signs of any problem. Knowing your vineyard literally vine by vine. You have to be observing all the time, even if you're doing the most mundane of tasks like cutting the grass.' Doing your vineyard work according to the phases of the moon, applying a dynamising spray of cow manure fermented and buried in a cow horn, using infusions of nettles, of comfrey, of horn silica for the vine's health – I've heard of these in vineyards all around the world. But the one thing that marks out all the biodynamic producers is an almost selfless devotion to their land and a positively military rigour in obeying the natural laws as laid down by biodynamics. And here in the loveliness of the Surrey Hills, against all the odds, the charming yet stubborn Nick Wenman is proving, once again, that you don't have to understand biodynamics to make it work, you just have to commit yourself to it and believe that, somehow, it *will* work.

Denbies Wine Estate

London Road, Dorking, Surrey RH5 6AA
www.denbies.co.uk Telephone 01306 876616
First planted 1986; vineyard area 107 hectares (265 acres);
total average annual production: approx. 500,000 bottles
At the winery: daily tours and tastings, groups by appt, 3 restaurants,
Vineyard Hotel, numerous events, weddings, Park Run every Saturday
and Bacchus marathon in September
Sales: cellar door, online, local and national stockists, export
Oz recommends Blanc de Blancs∗, Blanc de Noirs∗, Ranmore Hill
white and Noble Harvest

Once they got started at Denbies in 1986 the White family really set out to hit the ground running. They would plant 100 hectares (247 acres). From scratch. They would expect to harvest 10 tonnes of grapes per hectare – that's 1000 tonnes. Straight away. So a big winery would be needed. They'd better build it. A specially embossed, super-classy bottle would announce Denbies to the adoring world – they'd need a million of these. And they would buy six vans in Denbies' colours to rush the wine to all the customers panting with expectation. Pity no one thought about the wine. I remember the fanfare of the Denbies' arrival. I remember the staggering transformation of those bare slopes and meadows on the edge of the North Downs near Dorking into a thrilling vineyard reaching as far as the eye could see. And I remember the terrible disappointment on prising the cork from those self-confident bottles and thinking what a lot of fuss about nothing. The wine just wasn't any good. The fabulous potential was being wasted. So we all moved on to other wines.

That was the state that Chris White found things in at Denbies when he took over running the estate in 2001. Being close to London and having all those big winery buildings meant that it had been easy to build a tourist business – loads of coaches of indiscriminate punters all leaving clutching a bottle of something Denbies, probably on the sweet side, maybe not quite clean to the tongue. No matter. This tourism stuff was making Denbies a profit, but it was all being swallowed up by the wine side of the business making a whopping loss. Chris could see a fantastic future for wine tourism, of a much higher grade than the charabancs of the 1990s. But he needed a wine visionary to turn the vast potential of these thrilling vineyard sites into wine worth shouting about. An iconoclast, maybe? A talented, visionary 'bull in a china shop' Aussie, maybe? John Worontschak sounded like the perfect fit.

John had already made his mark in England with a series of fascinating wines from Stanlake, near Twyford in the Thames Valley. He was also the classic Aussie troubleshooter – heading off to countries like Russia, Brazil, Mexico, Uruguay and Peru and shaking up moribund wine cultures to produce something pretty decent which he then usually sold to British supermarkets. He'd been turning a sow's ear into a silk purse all over the world. So the Denbies challenge was perfect for him. Chaotic vineyards that he would need to tame and reorganise. Grapes from 20 or more different varieties, ranging from rubbish to excellent, and planted on everything from fabulous steep, protected south-facing chalk slopes to muddy frost traps down by the railway embankment. And he needed carte blanche. He saw the potential for some great wine, but he also saw the potential to make Denbies the UK's most important supplier of bargain-priced own-label wine to the supermarkets and wine clubs. And that's something he was an expert at. Any of you who have drunk own-label English wines from the supermarkets have probably been drinking a Denbies wine. And you've probably remarked that it was really quite delicious at a sub-£10 price.

When John arrived at Denbies as a consultant he found that they had been using a picking machine to harvest the grapes for years. As an Aussie he was used to that. As English vineyards proliferate, finding crews of skilled pickers for a big operation like Denbies is becoming a real challenge. The picking machine means that they can wait longer for ripeness, pick more quickly and pick more cheaply. And since 2017 a new vineyard manager has arrived who has implemented a scheme of minimum pruning for the vines. In some of the worst, most frost-prone parts of the vineyards, this has resulted in more bunches of smaller berries, with lower acid and no rot. And far bigger yields. One section of Müller-Thurgau was only yielding 2 tonnes, now it yields 25. The Reichensteiner plot was yielding 6 tonnes, now it's 60 tonnes. And that's the stuff going into your very tasty, own label English white.

That's the base of the business. But the best vineyards are being brought back from near neglect to being some of the best potential sites in southern England, and Denbies is aiming for these heights. Sites like Kit's Coty in Kent are already producing stunning Chardonnay still wine for Chapel Down. Denbies thinks they have a plot, at 40 per cent slope the steepest in England, to match that.

The sparkling wines have been beefed up and the basic wines are now good, while the top end is excellent, their 2013 Blanc de Blancs being one of the best from that vintage, and more recent releases are likely to be even better. And John now has free rein to experiment. He runs his own wine company, Litmus Wines, from Denbies, and it's mutually beneficial as John's desire to push boundaries benefits the style and quality at Denbies, too. Denbies carried out trials with Sauvignon Blanc in England, a work in progress, but just wait a few years. They have explored possibilities of super-dry sparklers with zero dosage (no sugar) cuvées. They have made a skin-contacted 'orange' wine and they regularly make England's best sweet wine from a single parcel of Ortega that always gets the mist settling on the vines in the autumn – just like Sauternes does in France. John is even pioneering wine in cans – and it's really good.

And Chris's tourist operation now brings 350,000 visitors a year, but at a much higher grade of customer. Early in 2019 he opened a carbon-neutral hotel next to the winery and by the summer it was already almost fully booked. Denbies can now produce a million bottles a year – of wine that people want to drink. 'These are wines I drink for pleasure,' said John. 'When we started people didn't drink English wine for pleasure very often.' They do now. It's taken over 30 years, but the Denbies dream is finally reaching fruition.

MORE VINEYARDS AND WINERIES

Biddenden Vineyards

Gribble Bridge Lane, Biddenden, Ashford, Kent TN27 8DF
www.biddendenvineyards.com Telephone 01580 291726
First planted 1969; vineyard area 10.5 hectares (25 acres);
total average annual production: 65,000 bottles
At the winery: self-guided tours and by appt, tastings, shop selling
Kentish serving platters as well as wines, ciders and juices, annual
food and drink day in June
Sales: cellar door (including ciders and juices), online,
local and national stockists
Oz recommends Gamay Noir and Gribble Bridge white

When I say that Biddenden Vineyards has been giving me pleasure for as long as I've been drinking English wine, the Barnes family would probably reply, yes, but we've been at it a good deal longer than that. In 2019, they celebrated 50 years of growing grapes and apples and making wine and cider at their bucolic farm on the wonderfully rustic Gribble Bridge Lane near Biddenden in

Kent. Apples were their main concern, but I come from apple country in Kent, and I've been listening to growers complaining about falling prices since I was a child. The family decided to diversify into grapes when Mrs Barnes heard a story about English vineyards being planted on BBC Radio 4's Woman's Hour. They initially planted the white Germanic varieties like Huxelrebe, Reichensteiner and Ortega that have now gone out of fashion. Well, not at Biddenden they haven't. Ortega is the flagship wine and is always a delight. They do make sparkling wine, and have some Pinot Noir, but they have a wonderful loyalty to the old-style grape varieties, mainly because wine drinkers love them. And there's one more Biddenden speciality I adore, but which they don't make very frequently – a fabulous, crunchy Gamay red that would sit proud with Beaujolais' best. Maybe with global warming they'll make it a bit more frequently.

Bluebell Vineyard Estates

Glenmore Farm, Sliders Lane, Furners Green, East Sussex TN22 3RU
www.bluebellvineyards.org Telephone 01825 791561
First planted 2005; vineyard area 40 hectares (100 acres);
total average annual production: 150,000 bottles
At the winery: self-guided tours, prebooked groups by appt, weddings, food festival, Join the Crush, blending workshops, private functions, food and wine pairings
Sales: cellar door; online, local and national stockists, export
Oz recommends Hindlip Classic Cuvée✷, Ashdown Estate White and Ortega

---

I must admit that one of the things that attracted me most about Bluebell was the fact that it's right next to the wonderful Bluebell Steam Railway. I was almost prepared to confess I found the gorgeous aroma of freshly burned coal smoke drifting through the wines. But that was a bit fanciful, even for me. Instead, the sparkling wines, made from Pinot Noir, Pinot Meunier and Chardonnay, but also from Seyval Blanc, have an uplifting purity of fruit, an orchard blossom scent, and fine, vibrant acidity which is very refreshing, while being quite different to the deeper, more creamy, nutty styles of most other Sussex sparkling wines. But

Kevin Sutherland, the winemaker, has never followed the pack, and that's particularly evident in his new range of 'Ashdown' still wines. They are unashamedly bone dry, but it works. The Ortega is mouthfilling and slightly tropical, the Bacchus is almost tart, but its bright elderflower scent makes it a delightful thirst quencher. These grapes are common in England. Chasselas isn't – it's a Swiss speciality and the Ashdown version is delicate and mild with a flavour of fluffy white apple flesh. And the rosé mixes Pinot Noir very successfully with Merlot, one of Bordeaux's leading red varieties. That's a British first. This is Kevin looking ahead to climate change. Merlot already ripens well enough in England to make tasty pink and it won't be long before it's making impressive reds here as well.

Davenport Vineyards

Limney Farm, Castle Hill, Rotherfield, East Sussex TN6 3RR
www.davenportvineyards.co.uk Telephone 01892 52380
First planted 1991; vineyard area 9.7 hectares (24 acres);
total average annual production: 29,000 bottles
At the winery: not open to visitors
Sales: online, local and national stockists
Oz recommends Limney Organic Sparkling*, Limney Vineyard
Horsmonden White and Diamond Fields Pinot Noir Précoce

It's very easy to decide that our damp and frequently cool maritime climate simply isn't suitable for an organic vineyard. Mildew and rot are perennial threats that have traditionally needed chemical sprays to combat them. Weeds grow like crazy if you're on remotely fertile soil. Global warming is bringing new pests along with the old, and it would be really nice to just spray them away.

But Will Davenport seems eternally unflappable as each new challenge arrives. He planted his first couple of hectares in 1991 and wasn't organic then, but by 2000 he had decided organic was the only way he wanted to farm, and to make wine for that matter. Over the years I haven't had much joy from English organic wine, and most English wine experts aren't sure it can work here. But Davenport's wines get better each vintage, led by the utterly delightful elderflower-scented Horsmonden White, made from a selection of the original plantings of Germanic early-ripening grape varieties that are still going strong. The wines are all now made with minimal intervention – natural wines? Pretty much. And his winery is carbon neutral, and won a sustainability award in 2018. But none of this would work if the wines didn't taste good.

Greyfriars Vineyard

The Hog's Back, Puttenham, Surrey GU3 1AG
www.greyfriarsvineyard.co.uk Telephone 01483 813712
First planted 1989; vineyard area 16 hectares (40 acres);
total average annual production: 75,000 bottles
At the winery: self-guided tours and by appt, groups summer
weekends, tastings, events, Young Wine evenings, honey from
hives on site
Sales: cellar door, online, local and national stockists, export
Oz recommends Classic Cuvée✳, Cuvée Royale✳ and Blanc de
Noirs✳

There must be something in the water at my college. Mike and Hilary Wagstaff who own Greyfriars were both at Pembroke College, Oxford. So was Nicholas Coates of Coates & Seely, and several other wine tyros. Or maybe we all just spent rather a lot of time in the college bar.

Greyfriars was actually established just below the Hog's Back chalk ridge outside Guildford in 1989, but the Wagstaffs swept in in 2010 full of ideas and ambitions which they are triumphantly realising. They expanded the vineyards to 16 hectares (40 acres), most of them steep and south-facing which not only maximises the chalk's suitability for growing vines, but also provides brilliant, cool, slightly damp conditions for ageing sparkling wines, so they have dug cellars straight into the chalk that will house 250,000 bottles. Their sparkling wine immediately drew attention because of their use of oak barrels – commonplace now but less so then. And they've planted some Sauvignon Blanc. Experts said it wouldn't ripen here, but it does. And they know that just making wine is to miss a trick. People want an 'experience'. More and more wineries are doing 'meet the winemaker' events. Greyfriars go one better – and hold 'young wine' evenings where you can taste all the base wines that go into the different blends, then the pre-release sparklers and then the sparklers that are mature enough for sale – in other words, the whole path. I used to have to go to Champagne for a similar experience.

Nutbourne Vineyards

Gay Street, Pulborough, West Sussex RH20 2HH
www.nutbournevineyards.com Telephone 01798 815196
First planted 1980; vineyard area 10.5 hectares (26 acres);
total average annual production 40,000 bottles
At the winery: open May–Oct, guided tours by appt, tastings
(no appt needed), lunch and supper for groups, family of alpacas,
historic windmill, private parties and weddings
Sales: cellar door, order over the phone or via the 4 restaurants
in London, local and national stockists
Oz recommends Sussex Reserve✳ and Nutty Brut✳

It may have been because every time I had a Nutbourne wine my mind filled with visions of Tina Turner pumping out 'Nutbush City Limits'. Or it may have been that at the end of the last century Nutbourne wines were reliably fruity and scented and clean when most of their rivals were not. Either way I always thought Nutbourne was a bit special. But it was only when I crawled down a tiny lane in Sussex searching for the more famous Nyetimber that I came upon the real reason. Nutbourne is on the same hidden lane and the vineyards are on the same south-facing, protected, warm greensand soils. Now all that fruit and perfume made sense.

Yet while Nyetimber has pursued a glittering international path, Nutbourne is irresistibly local in its claims. Well, not quite. In one way it's ahead of its time. Peter Gladwin, the owner, is a successful restaurateur. His sons, Oliver and Richard, run four restaurants in London which specialise in local Sussex produce and Nutbourne wines are right at the front of the wine list. 'Farm to Fork' is an increasingly common concept for food and Nutbourne has gone one further and added 'Grape to Glass'.

Rathfinny Wine Estate

Alfriston, East Sussex BN26 5TU
www.rathfinnyestate.com Telephone 01323 871031
First planted 2012; vineyard area 91 hectares (226 acres);
total average annual production: 97,000 bottles
At the winery: tours by appt, tastings, Tasting Room restaurant, Flint
Barns Dining Room, dining events and corporate events, Flint Barns,
both B & B and exclusive use, Rathfinny Trail footpath across the
South Downs
Sales: cellar door, local and national stockists, export
Oz recommends Blanc de Noirs✳ and Cradle Valley White

I don't think there's ever been a winery that has made so much noise and attracted so much attention before any wine had even been made than Rathfinny. This is a grand, ambitious project which Mark and Sarah Driver bought as a 243-hectare (600-acre) arable farm – with no vines – in 2010. They didn't plant vines until 2012 but by then they had already been the subject of a massive national newspaper feature about what they were going to do with these challenging windy Sussex acres and the PR has rolled on ever since.

Well, Rathfinny is up and running, but it hasn't been an easy ride. The chalk slopes near Eastbourne do face south, but they do get a belting from the winds coming off the sea only a couple of miles away. The vines planted so far have needed 10,000 trees to be planted as windbreaks. Luckily Rathfinny has a very plucky New Zealand vineyard manager and the winemaker is a pretty serious Frenchman whose first sparkling vintages from 2014 and '15 grapes were elegant and restrained, as though they were testing the waters.

A still wine under the Cradle Valley label from Pinot Blanc and Pinot Gris is more evidence that these two varieties should have a great future in England. Mark Driver is very keen for his county to have a separate PDO, or Protected Designation of Origin, so that wine drinkers around the world will start asking for a 'Sussex' wine rather than just an English wine. Well, this approval worked for places like Marlborough in New Zealand, so let's see.

Stopham Estate Vineyard

Stopham, Pulborough, West Sussex RH20 1EE
www.stophamvineyard.co.uk Telephone 01798 865666
First planted 2007; vineyard area 6 hectares (16 acres);
total average annual production: 30,000 bottles
At the winery: tours by appt, drop-in for free tastings,
new tasting room and shop, Stopham Festival in Sept
Sales: cellar door, online, local and national stockists, export
Oz recommends Pinot Blanc and Pinot Gris

I first stepped onto the Stopham vineyard in August 2012. It says something about village life in Sussex that Simon Woodhead, the boss, had found me within five minutes, having been told, hey, there's some bloke wandering about in your vines. The very first thing I had noticed apart, from the fabulous view over the River Arun to the South Downs, was how incredibly sandy the soils were. They are part of the same greensand south-facing ridge which has made Nyetimber famous. The second thing I noticed was that there were almost no grapes on the vines, and Simon stoically explained the tribulations of establishing a vineyard in England. He planted the vines in 2007. In 2010 frost hit the young vines, but he still made 10,000 bottles of wine. In 2011 he had rubbish weather at flowering and only managed 9000 bottles. In 2012 at flowering he thought he would make 30,000 bottles but the summer never came in 2012, and he only made 3000 bottles.

This is the kind of story you will hear a lot from British vineyard owners, but they almost all think it is worth it. Simon was a designer for the McLaren Formula One team. Even if he doesn't always get the crop he wants, the pace of life in a Sussex vineyard massively increases his life expectancy. And he did a very clever thing. He saw that even if Chardonnay and Pinot Noir were difficult to ripen for still wines, Pinot Blanc and Pinot Gris could be the future for superb, balanced still whites. I have just opened the Stopham Pinot Blanc 2011 to see how it was getting on – delicious. It could have been a Chablis.

Tillingham

Dew Farm, Dew Lane, Peasmarsh, East Sussex TN31 6XD
www.tillingham.com Telephone 01798 263123
First planted 2018; vineyard area 8 hectares (20 acres);
total average annual production: 50,000 bottles
At the winery: self-guided and guided tours and tastings, restaurant,
wine bar, wine shop, 11 rooms accommodation plus a range of bell
tents and lodges, courses (yoga, baking, foraging, arts and crafts),
private hire, parties and receptions
Sales: cellar door, online, local and national stockists, export
Oz recommends Rosé, Qvevri White and 'R' Red

I knew nothing about Tillingham until I went to the Real Wine Fair in 2019. As soon as I found the English section, I certainly wasn't going to miss Ben Walgate, the owner and winemaker – a blond bombshell bundle of energy and ideas and enthusiasm. And that was before I saw the labels for the wines – wonderfully expressive, imaginative and I don't think a single one actually focused on the grape variety or even mentions Tillingham. I'm not sure most of them even mentioned the vintage.

And then I tasted the wines. They were all made from bought-in grapes because the vineyard, at Peasmarsh looking out toward Rye in East Sussex, was only planted in 2018, but the wines exhibited an originality and a conviction that will make Tillingham a star of the future. The objective here is to use organic and, eventually, biodynamic grapes, and the objective is to use low to minimal intervention – 'natural' winemaking. One thing all great winemakers must have is a vision of flavour and Ben Walgate certainly has that – from his cloudy but vibrant 'Pet Nat' rosé bubbly, to his wines made in Georgian terracotta *qvevri* jars. And just so that you know he's not a geek or an ideologue, he also makes cider in these jars – a delicate and delicious cider that has straight away become one of Kent's best. No one knows what the future path of English wine will be, but there's no doubt that if biodynamic vineyards and 'natural' winemaking are to play a part, Ben Walgate will be leading the charge.

Wiston Estate

Wiston Estate Winery, North Farm, London Road, Washington,
West Sussex RH20 4BB
www.wistonestate.com Telephone 01903 877845
First planted 2006; vineyard area 6.5 hectares (16 acres);
total average annual production: 40,000 bottles
At the winery: tours and tastings by appt, wine dinners, weddings,
South Downs Way conservation area
Sales: cellar door, online, local and national stockists, export
Oz recommends Wiston Estate Cuvée✳ and Blanc de Noirs✳

Wiston Estate's vineyards have a long and noble heritage – they are part of the 2428-hectare (6000-acre) farming estate spread across the South Downs near Worthing that the Goring family has been running since 1743. The winery is a bit less noble – it's an old turkey shed and when I first visited it, you could virtually see the feathers stuck to the walls. But bang in the middle was a truly impressive, wooden Champagne press – a Coquard – of the type that all the great traditional Champagne producers have. But almost no one else – and certainly no one else in England – has one.

It took Pip Goring, the South African wife of Harry Goring, 34 years to persuade him to plant vines, in 2006. It took a single meeting to persuade Irishman Dermot Sugrue, already a star performer at Nyetimber, to join them as winemaker. He saw the massive potential of a protected South Downs chalk site, and has since made a string of wines that are remarkable for their incisive flavours and their veiled suggestion of something slightly wild being held in check. That must be the Celtic influence of Dermot, who also makes his own private label at Wiston called Sugrue Pierre, a challenging but memorable sparkler aimed right at the top of the quality tree.

WESSEX

1. Furleigh
2. Bride Valley
3. Langham
4. Leckford Estate
5. Coates & Seely
6. Hattingley Valley
7. Jenkyn Place
8. Exton Park
9. Hambledon

WILTSHIRE

HAMPSHIRE

•5

•7

•4

•6

● Winchester

•8 •9

Southampton ●

DORSET

•1 •2 •3

● Dorchester

Wessex as a wine region is strongly defined by the wealth of chalk and Jurassic limestone that covers much of Hampshire, Wiltshire and Dorset. Hambledon Vineyard is given credit for beginning the revival of English vineyards in 1951 and this site is right in the heart of the rolling chalk hills of the South Downs, just to the north of Portsmouth. Other top-quality vineyards and wineries have opened to immediate success in a wide swathe that goes north across Hampshire and south to the New Forest and the Isle of Wight. But a real core of high quality has developed around Winchester, spreading north from Hambledon through Raimes and Hattingley Valley, near Alresford, as far as Coates & Seely west of Basingstoke. And I'm sure there will be many more to come in the near future. Wiltshire has good soils for vines but for the most part is a little high and often a bit exposed. Dorset can be rather high and open to the winds and rains of the English Channel, but there is a wealth of good soil and, when sufficiently protected, exciting wines have already been made from areas such as Dorchester (Langham) and Bridport (Bride Valley and Furleigh).

Exton Park Vineyard

Allens Farm Lane, Exton, Southampton, Hampshire SO32 3NW
www.extonparkvineyard.com Telephone 01489 878788
First planted 2003; vineyard area 24 hectares (60 acres);
total average annual production: approx. 70,000 bottles
At the winery: visits by appt, tours, tastings, events
Sales: online, local and national stockists, export
Oz recommends Brut Reserve*, Rosé*, Blanc de Noirs* and Rosé
Meunier*

'I'm not here to make a copy of Champagne. I want people to understand that we are here to make a new category of wine. Winemaking in England is an adventure and a challenge. We are just at the beginning of the journey.'

I'll drink to that. These are the kind of sentiments to inspire a whole generation of young English winemakers to reach for the stars. But it's not an Englishman who's uttering them. It's a French woman. A woman who has made some of the most famous wines in France and now is one of the most passionate believers in England and its ability to make the greatest sparkling wines in the world.

Corinne Seely is the winemaker at Exton Park, high on the chalk slopes of the South Downs in Hampshire. She's not your usual sparkling winemaker. She used to make the rip-roaring red wines at Bordeaux's Château Lynch-Bages and Domaine de Chevalier, and she's done her time in the sun-drenched vineyards of Portugal's Douro Valley and Australia, too. But you could not find a more passionate exponent of the specialness of the fruit she can find growing in England's cool, chalky slopes, nor someone more determined to turn it into wines unlike any other.

Luckily, she's also got the perfect foil. Fred Langdale, who is in charge of the vineyards, is as quintessentially English as she is intriguingly French, but he's also done his time in New Zealand and South Africa, as well as handling Nyetimber's vines in Sussex. It's not often you feel such a close bond between vineyard manager and winemaker, but you do here. And it's vital. If Corinne is determined to make wines unlike any others in England, and definitely unlike any in France, she needs to have a supply of very particular grapes.

Exton Park is surely the place to supply them. These are high, exposed vineyards, with barely any soil before the vine roots hit a bed of chalk many metres deep and as blindingly white as any I can remember. 'Is it natural for vines to grow on such pure chalk?' I asked Fred. 'If they didn't like it,' he replied, 'they wouldn't grow. Onions don't.' That's put me in my place. We were standing on the shoulder of the vineyard looking across the valley to Old Winchester Hill. It's a view of heart-stopping beauty, in a county of wonderful views. Such beauty demands that the wine reflects it.

The first time I tasted Exton Park, I was astonished. The wines had a purity, a clarity, but also an intensity of flavour and a uniqueness of scent I'd never come across before. It was as though the very essence of the grape, that essence which is often lost or at best translated into something entirely different, had been preserved here; the cool breeze, the scent of wild flowers, the promise of hillside herbs barely dried by our northern sun, and the chilly mineral core of the damp but blazing white chalky soil – they're still there in the wine.

That's the whole point at Exton Park. The winery has been situated at the heart of the vineyard. Corinne says it would be insane to lose this high downland freshness that she finds in her grapes. So she says the target she sets for the time from being picked off the vine to the grape press is 5 minutes flat. And when the grapes reach the press she uses a rare kind of system that covers the grapes in nitrogen so that oxygen can't begin to degrade them. This also means she can press the grapes very slowly and for much longer than usual. Which has Fred tearing his hair out as he calculates the cost of his expensive team of pickers waiting in vain for the next call for grapes. But the results are worth the wait.

And Corinne is also unusual in that she barely makes any wine with a vintage date on it. Even with global warming, our climate simply can't be be relied on year by year, so almost all her wine is 'non-vintage'. Other wineries do this, but usually base their non-vintage on a single harvest – say, 80 per cent – and then add 20 per cent of older, mellower wine. At Exton, they've now built up sufficient reserve in the tanks that there's no such thing as a 'base' vintage. Corinne looks at all the wines right back to about 2014 and simply chooses the blend that she thinks will interpret her precious vineyard most truly. The wines are wonderful now. Just think what kind of character she'll be getting in her blend in 10 years' time.

Hambledon Vineyard

The Vineyard, East Street, Hambledon, Hampshire PO7 4RY
www.hambledonvineyard.co.uk Telephone 02392 632358
First planted 1952 (under different ownership); vineyard area
currently 97 hectares (240 acres); total average annual production:
approx. 140,000 bottles
At the winery: open all year Mon–Fri, groups by appt, guided tours
Sales: cellar door, online, local and national stockists, export
Oz recommends Classic Cuvée✳, Rosé✳ and Première Cuvée✳

Hampshire is full of chalk enthusiasts, but they don't come any more enthusiastic than Ian Kellett at Hambledon Vineyard. Just having chalk that seems to be the same as that of Champagne isn't good enough for him. The most celebrated chalk soils of Champagne are from a very precisely delineated cluster of sites on the 'Côte des Blancs' near Epernay – the short, sweet, east-north-east-facing outcrop of chalk barely 18 kilometres (11 miles) long that grows all the greatest Chardonnay grapes in Champagne. And the sweet spot, in the heart of the Côte, encompassing villages like Le Mesnil where the top wines sell for hundreds of pounds a bottle – well, that identical chalk is the chalk at Hambledon Vineyard.

You could argue with him, but Ian is a biochemist and very sure of his facts. You could say, hang on, why are the best Chardonnay vineyards in Champagne facing away from the warm afternoon sun, and Ian would counter that almost none of the great Champagne vineyards, for Pinot Noir or for Chardonnay, face toward the warm afternoon sun. And nor do his Hambledon vines. They face south-east, toward the cooler morning sun. Ian tells me that the cooler morning sun has greater photosynthetic powers than the warm afternoon sun and so creates more ripeness and

This is where it all started – the manor house at Hambledon and the original vineyard below, now producing superb Chardonnay grapes.

sugar in the grapes. And yet the acids stay higher because of the lack of warmth. And acid may not sound enticing – but it is fundamental to making fresh, appetising sparkling wine. By the look of their vineyards, Champagne has been chasing fresh acid, not bulging sugar levels, for hundreds of years. Ian Kellett at Hambledon has learnt the lesson.

And Hambledon is a fascinating place for him to put his theories to the test. This is the site of the first commercial vineyard to be planted in Britain since 1875 – it was 1952 when a retired general, Sir Guy Salisbury-Jones, stood at the window of his house on Windmill Down and gazed across the steep chalky slope beneath him and thought that this could be a vineyard. Hopefully he also looked upward to the brow of the hill and took note of the Bat and Ball pub, because that's where cricket was invented.

In those long gone days, England didn't play against Australia or South Africa – they played against Hambledon. The landlord of the Bat and Ball pub wrote the rule book for cricket. A chap called Thomas Lord took it up to St John's Wood in London, but as you stand at the top of the slope where our first commercial vineyard in modern times was planted, you can also see where cricket began – just up there on the ridge.

Ian is determined to put English sparkling wine at the head of the world's bubbly. He reckons Hambledon already has more sunshine than Reims does, in the heart of Champagne. He reckons that his vineyards are only 5 per cent cooler than those of Chablis in northern Burgundy. He does tub thump for Britain, but he is very aware that the French have been excelling at fizz for a lot longer than us. His chief winemaker, Hervé Jestin, comes from Vertus in Champagne's Côte des Blancs. His consultant, Didier Pierson, comes from the neighbouring village of Avize. As Ian says, we have the greatest sparkling wine skill set in the world in Champagne, and they're keen to help. And he's keen to meet them halfway. He's learnt French.

Hambledon's wines will soon be coming from over 81 hectares (200 acres) of pale chalky soil, all on the estate. The wines are already made with what are possibly the gentlest, gravity-fed methods in the UK, and the finished wines will soon be stored in a new cellar dug into the damp, cool, chalk soil, which hardly needs air conditioning even in high summer.

And the wines don't taste like those of their neighbours. Nor do they taste like anything I know from Champagne's Côte des Blancs. The whites create a striking shotgun marriage between high acidity, loft-stored apples and a mellow coating of honey. The pink is fresh, bright, exuberant and fruity. And they all have an intriguing lick of salt. When you stand at the top of that south-east-facing slope by the house, the sea is only 8 kilometres (5 miles) away.

Hattingley Valley Wines

Wield Yard, Lower Wield, Alresford, Hampshire SO24 9AJ
www.hattingleyvalley.com Telephone 01256 389188
First planted 2008; vineyard area 10.5 hectares (26 acres) plus
13 hectares (32 acres) managed; total average annual production:
approx. 300,000 bottles
At the winery: open all year Mon–Sat (excluding harvest),
private groups welcome, guided tours by appt, tastings including

masterclasses in sparkling wine, corporate and private events, catering on request
Sales: cellar door, online, local and national stockists, export
Oz recommends Rosé*, *Blanc de Blancs**, *Classic Reserve** *and Entice (sweet Bacchus)*

I'd been hearing about Hattingley Valley for years before its wines first hit the shelves. Actually, before its wines had even been made. In fact, before the vines had even been planted. I'd known Simon Robinson for years in London, where he was an extremely successful lawyer. We would taste wine together, and he would tell me about his farm in Hampshire with its lovely south-facing slopes on deep, pure downland chalk. And we'd josh together about why he hadn't got a vineyard, and when was he going to swap his pinstripe suit for Wellington boots and dungarees and start planting vines.

And one year – it was 2008 – he said, 'I've done it. I've started planting.' There was one particularly perfect slope which ran along the side of Chalky Hill Road. He didn't need a geologist to tell him – this was where he should plant his Chardonnay, Pinot Noir and Pinot Meunier. He also put in a few rows of Bacchus and Pinot Gris. And then, in a fit of devilment, some Chenin Blanc, a white variety from the Loire Valley – but that's long gone, it never ripened beyond frozen pea state. The other grapes did, though, and yet another great vineyard site on Hampshire's billowing chalk had been identified – along with the fabulous view across the meadows, hedgerows and copses of ancient woodland, plus the odd church spire that seems to be so common in this beautiful South Downs wineland.

But just growing grapes was never going to be enough for a powerhouse like Simon. By 2010 he had built a seriously impressive but sensitively designed winery a mile or so from the vineyard, and just close enough to the village pub for those moments after a long day when the thirst gets to you. Simon may delight in being a Hampshire vineyard owner, but his ambitions were much broader than that. He wanted to actually make the wine and make it in his own winery. He'd need help. Well, he found the right person. Emma Rice was doing a fair bit of freelance custom-crushing and winemaking and obviously had wine in her blood because her mother would take her on grape-picking expeditions to Childsown vineyard near Chichester. I've had the 1976 – probably picked by Emma's mum – when Emma wasn't even a toddler. She had a wide perspective on wine. She'd been a wine editor in London, she'd worked in New Zealand, Australia and California's Napa Valley and was now absolutely ready for a challenge closer to home.

So Simon asked her to build his winery. Sure she would, but then she asked him would he also want to make wine for other people? Exactly the right question for Simon. And so the modern Hattingley Valley was born. Above all, Emma makes tremendous sparkling wines under the Hattingley Valley label, which have twice won her the title of 'Winemaker of the Year' in Britain. One of the features that immediately made her wines stand out for their sumptuous texture and flavour as well as their fabulously focused acidity and caressing foam, has been the use of oak barrels to ferment and age a proportion of her base wines. Lots of winemakers do it now, but few as well as Emma, and in her barrel hall she has 160 different oak barrels of various ages to choose from.

And if you haven't tasted Hattingley Valley wine, you may well have tasted Emma Rice wine, because half of the Hattingley business plan is about making wine for other vineyards. And they win awards: Cottonworth, Alder Ridge, Roebuck, Raimes – all gold medal winners, all with their own vineyards, all the wines made by Emma Rice at Hattingley Valley.

'Do you want to see my new shed?' asked Emma. Some shed. After planning to use 430 tonnes of grapes in 2018, but receiving 680 tonnes, she needed a new shed for all the bottles she'd made. In 2019, the new 'shed' appeared – a gentle giant in brick and pine capable of storing and processing 870,000 bottles of tiptop sparkling wine. 'I didn't think I'd ever be able to do this when I started,' said Simon. 'Nor me,' added Emma. But they did.

MORE VINEYARDS AND WINERIES

Bride Valley Vineyard

Court House, Litton Cheney, Dorchester, Dorset DT2 9AW
www.bridevalleyvineyard.com Telephone 01308 482767
First planted 2009; vineyard area 10 hectares (24½ acres);
total average annual production: 19,500 bottles
At the vineyard: guided tours and tastings during the summer; private
ones by appt
Sales: cellar door, local and national stockists, export
Oz recommends Brut Reserve✳ and Blanc de Blancs✳

'I can't imagine a better or more satisfying culmination to my
career in wine than this.' For most English vineyard owners,
the vineyard is their first leap into the world of wine. For Steven
Spurrier, it provides a glittering climax to a life during which he
has done almost everything else – from writer to seller, from judge
to teacher – and he's done it all over the world. But this just might
be his best venture of all. His wife Bella bought a house in Dorset
at the end of the 1980s and then bought 85 hectares (210 acres) of
land to raise sheep. It may not have been a great success as a sheep
farm, but the soil was mostly fabulous chalk and one slope looked
enticingly toward the west and the warm afternoon sun. With global
warming showing what England could do at estates like Nyetimber
and Ridgeview, well, what would you do if you already owned what
looked a perfect site for Chardonnay and Pinot? Of course you would.
So would I. You plant the vines. Steven started in 2009 and made the
first Bride Valley wine in 2011. The site has proved to be erratic for
yields, but excellent for quality, especially Chardonnay. Steven says
that he wants a wine of lightness and elegance – an aperitif wine.
That's exactly what his Blanc de Blancs epitomises.

Coates & Seely

Wooldings Vineyard, The Harroway, Whitchurch,
Hampshire RG28 7QT
www.coatesandseely.com Telephone 01256 892220
First planted 2009; vineyard area 12 hectares (30 acres);

total average annual production: 60,000 bottles
At the winery: tours by appt
Sales: cellar door, online
Oz recommends La Perfide✳ and Rosé✳

You probably have to live in France and with the French to be able to get away with scattering jokes about the French all over your wine labels. Nicholas Coates and Christian Seely are two of the most English, most urbane chaps you could ever hope to meet, but Christian's day job is in France, running the wine estates for the giant AXA insurance group – which include such superstars as Château Pichon-Longueville in Bordeaux. So when he labels his English sparkling wine as 'Britagne' he expects to get a laugh from the French. As he does with his description of the Champagne method as 'Méthode Britannique', and he certainly gets a response by labelling his top wine as La Perfide – the French have called England 'Perfidious Albion' for centuries.

But jokes apart, Coates & Seely is consistently high quality, with grapes grown on pure Hampshire chalk. Yet Christian has a deep respect for the French and all the equipment, including a range of revolutionary concrete eggs to replace wooden barrels (manufactured by a company specialising in concrete mausoleums) – as well as the winemaker – they're all French. But the wines are no mere Champagne imitations – simply beautiful sparklers whose long-term objective is to be the best in the world, and English.

Furleigh Estate

Furleigh Farm, Salway Ash, Bridport, Dorset DT6 5JF
www.furleighestate.co.uk Telephone 01308 488991
First planted 2005; vineyard area 16 hectares (40 acres);
total average annual production: 50,000 bottles
At the winery: guided tours every week, groups by appt, tastings (no need to book), events including local chef pop-ups, Nordic walking, wine club
Sales: cellar door, online, local and national stockists, export
Oz recommends Classic Cuvée✳ and Rosé✳

You would have expected Ian and Rebecca Edwards to know better. They had both made successful careers as actuaries in the pensions business. Surely they would know that starting a vineyard and winery with your retirement money was a risk they would never have professionally recommended to anybody else.

But wine gets you like that. The idea of starting a vineyard gets under your skin. Many Brits dream of it, then talk to financial experts like Ian and Rebecca and decide not to be so silly. Well, neither of them seems to have tried that hard to talk the other out of it. And luck strolled by. The dairy farm near Bridport in Dorset where Rebecca had grown up came back on the market. They had money in their pocket. And they thought that one patch of south-facing land on the farm looked just perfect for a vineyard, so they bought it. They planted the first vines in 2005 and built a winery in 2007 while Ian went off to Plumpton College to learn about winemaking. He was obviously a good student. I didn't come across Furleigh fizz until the International Wine Challenge in 2013, when I was dumbstruck by the 2009's fabulous creamy, bread-crust savoury richness, its nutty depth balanced by vivid, bracing acidity – and this was Furleigh's first wine. Grapes ripen a little later in Dorset than further east in Hampshire, but the Furleigh vineyard is well protected against the winds from the English Channel just a few miles to the south as the great cliffs of the Jurassic Coast dip down briefly around Bridport. Furleigh also makes superb sparkling rosé – another Trophy winner at the International Wine Challenge – as well as some still Bacchus and Rondo.

And I always thought the wines would age well. Obviously Ian and Rebecca thought so too. In 2019 they launched a 'super-premium' range of sparklers called 'From the Oenotheque' of specially aged wine. The first release was a 2010 and I'm sure it's got another ten years ahead of it.

Jenkyn Place Vineyard

Jenkyn Place, Hole Lane, Bentley, Hampshire GU10 5LU
www.jenkynplace.com Telephone 01420 481581
First planted 2004; vineyard area 5.5 hectares (13½ acres);
total average annual production: 20,000 bottles

At the vineyard: tours and tastings by appt, tastings, events
Sales: cellar door by appt, online, local and national stockists
Oz recommends Classic Cuvée✳ and Blanc de Noirs✳

I sometimes think that Jenkyn Place's wines reflect their owner, Simon Bladon, a hard-nosed property developer from Yorkshire who had no interest in vineyards when he bought Jenkyn Place as a family home. But the house was surrounded by abandoned hop gardens, the poles that were once smothered with green, fragrant hops now angular, forlorn and ugly. They had to go, and if Simon hadn't tasted Nyetimber's wines and been astonished by their style and flavour, there would still be no Jenkyn Place wine.

And that would be a pity. The estate is beautifully located on the Hampshire chalk. Having been thrilled by Nyetimber, Simon managed to persuade Dermot Sugrue, the ex-Nyetimber winemaker, to create the wines for Jenkyn Place. Dermot is Celtic, fiery and opinionated. Simon is stern, earthy and opinionated. Between them they have developed some fabulous, cool, often lean, rarely seductive, sparklers, usually requiring time to age and blossom. And I'm sure any Yorkshireman would say that sounds just right.

Langham Wine Estate

Crawthorne Farm, Crawthorne, Dorchester, Dorset DT2 7NG
www.langhamwine.co.uk Telephone 01258 839095
First planted 2009; vineyard area 11 hectares (27 acres);
total average annual production: 50,000 bottles
At the winery: tours (guided, self-guided and private groups), tastings,
the Vineyard café, events
Sales: cellar door, online, local and national stockists, export
Oz recommends Blanc de Blancs✳ and Brut Reserve✳

Langham rather crept up on me. It's actually the largest single-site vineyard in the south-west of England. Three Choirs Vineyards in Gloucestershire is much bigger but is spread over several sites. At 11 hectares (27 acres) of vines, Langham is dwarfed by many other vineyards in Wessex, the South-East and East Anglia. And it isn't that old. Justin Langham only expanded his father's hobby vineyard into an ambitious site for making fine sparkling wines out of chalk-grown fruit in 2009.

But it wasn't until 2019 that I really worked Langham out – and then I got hit between the eyes. At the WineGB Awards there was a plethora of superb sparklers, often complex, rich and fascinating, but one simply shone for its purity, its focus, its electric acidity and uplifting style – Langham Blanc de Blancs. And it won Best Wine in Show. In fact, Langham's other wines are quite different in style, being wilder and pushing back boundaries – delicious but more challenging. And this makes sense. Sparkling winemaking is often a quite technical effort, but at Langham the winemaker, Daniel Ham, is intent on making wines with as little technical interference as possible, using wild yeast ferments whenever he can, and cutting sulphur as low as he dares. In all this, he's following the current eco-friendly mood in a way that few winemakers are able to do. And it's obviously working.

Waitrose/Leckford Estate

The Estate Office, Leckford Estate, Stockbridge,
Hampshire SO20 6DA
www.leckfordestate.co.uk Telephone 01264 812110
First planted 2009; vineyard area 4.7 hectares (11½ acres);
total average annual production: 30,000 bottles
At the winery: not open to visitors
Sales: farm shop, online via Waitrose Cellar, selected branches of
Waitrose and Partners
Oz recommends Leckford Estate Brut ✳

Great Britain is lucky to have Waitrose. This is a major supermarket group with almost 400 major stores and lesser outlets, and with an enviable reputation for top quality at a fair but not discounted price. But there's more. Waitrose own a vineyard, their very own Leckford Estate in the Test Valley bang in the middle of the action that is making Hampshire, with its endless rolling slopes of downland chalk, such an exciting new English wine region.

Waitrose has had a farm at Leckford since 1928 – 1620 hectares (4003 acres) of arable, grazing and horticultural land which supplies significant amounts of produce to Waitrose stores. But the vineyard is new – planted in 2009 with the Champagne grapes of Chardonnay, Pinot Noir and Pinot Meunier.

This was a bit of a gamble, because, although there had been a considerable spike in British vineyards since the hot summer of 2003, no one was quite sure yet whether you could regularly ripen the Champagne varieties in the UK, particularly out on the exposed chalk slopes of somewhere like Leckford. But Waitrose had gambled before on English wine. To be honest, being one of the smaller supermarkets they had realised early on that they needed to keep surprising their customers with unexpected delights. They were the first supermarket to bring in wines from New Zealand and South Africa and they realised that own-label Champagne could be produced that was better – and cheaper – than most well known labels. And way back in the 20th century they took an almost perverse decision to major on English and Welsh wines.

And it has paid off handsomely. By most reckoning Waitrose sells between 60 and 70 per cent of all English wines offered on the High Street and through supermarkets. In over 300 of their stores you can always find four sparkling and four still wines. They will be from the big producers, and from Leckford. But the reason the winemakers of England and Wales must thank Waitrose is that in vineyard areas, the local stores stock the local wines. You want to try Gwin Y Fro from Cowbridge in Glamorgan, three local stores have it. Keen on sampling Renishaw Hall from Derbyshire? Eighteen local stores stock it. Sixteen Ridges from Herefordshire? Ten local stores stock it. Smith & Evans from Somerset? Five local stores. Maud Heath from near Chippenham in Wiltshire? Six local stores. Altogether Waitrose has about 105 different wines, still and sparkling, from 48 producers in 20 counties. This is a fantastic route to market which small producers could otherwise never identify or afford. And if you live in Norfolk but feel like drinking something Welsh or live in Wales and pine for something bracing from Norfolk, the Waitrose Cellar website offers over 90 per cent of the whole range. Obviously including Leckford Estate.

1. Polgoon
2. Camel Valley
3. Sharpham
4. Lyme Bay
5. Three Choirs

SOUTH-WEST

This is a large sprawling region. It starts in the north-west with
Herefordshire and Worcestershire, the land of cider, and then
continues south through Gloucestershire, stretching from the open
limestone plateau of the Cotswolds right across the River Severn
to the Welsh border, through Bristol and Bath and into Somerset
with its widely varying sites from the high Mendips down to the
flatlands of the Somerset levels. And finally there's the broad spread
of Devon and the more challenging final destination of Cornwall
(and on into the Isles of Scilly out in the Atlantic where there is a
vineyard or two). There is absolutely no stylistic similarity between
the two extremes of the region, but lots of local variants are
popping up as sites are identified that are a little warmer than the
norm, and, because the prevailing westerly winds often bring rain,
a little drier too. Soils are more varied than in almost any other
wine region in Britain. So far, many of the best wines have come
from grapes like Seyval Blanc and Bacchus rather than the current

favourites of Chardonnay and Pinot Noir, but sites as far apart as the Cotswolds (Woodchester Valley) and Cornwall (Polgoon) have also been successful with surprise varieties like Sauvignon Blanc.

Camel Valley

Little Denby Farm, Nanstallon, Bodmin, Cornwall PL30 5LG
www.camelvalley.com Telephone 01208 77959
First planted 1988; vineyard area 10 hectares (24½ acres);
total average annual production: approx. 100,000 bottles
At the winery: tours and tastings, groups by appt,
2 self-catering cottages
Sales: cellar door, online, local and national stockists, export
Oz recommends Pinot Noir Rosé Brut❋, Brut❋ and Bacchus

For a winery situated almost at the extremity of English wine activity Camel Valley has exerted a quite undue influence on our national wine scene. The winery and its compact vineyard are snuggled into the Camel River Valley on the north Cornish coast near Bodmin. I know that coast. When it's sunny, it's beautifully fresh, but the weather can change in moments as rain sweeps in like a grey curtain. And you rarely forget that the forbidding uplands of Bodmin Moor are barely a mile or two from the vines. So the steep, south-facing protected slope that Camel Valley has planted is a precious little paradise in a rough and tumble part of England. But that's not the reason Camel is famous.

The Lindo family is why Camel is famous. Father Bob, mother Annie and son Sam, but above all, Bob. He's a force of nature who shouldn't really be here. Having a head-on collision in a fighter jet is not reckoned to do much for your chances of a long and contented retirement. But Bob Lindo, scarcely credibly, survived such a crash when he was an RAF fighter pilot, and it's almost as if the experience made him vow to never waste another moment. The Lindos had already bought a compact, 33-hectare (81-acre) farm on the slopes of the River Camel for not much money since no traditional farmer could ever make a living out of it. This was to be the focus of their contented retirement, probably running a few sheep. But Bob was only 38 years old when he had to retire from the RAF. And one thing stuck in his mind about their farm: the slope beneath the house was hot. And it was very steep and it was dry. So in 1988 the Lindos planted vines.

They started winning awards with their very first wine – a Seyval Blanc from their warm, dry slope. And Bob quickly seemed to assume the mantle of mouthpiece for the rapidly growing band of English winemakers, always putting himself forward when opinions or actions were needed. Memorably, in 2012 when he discovered that Gatwick Airport was running a big Olympic promotion and he found that not only was there no English wine featured, but that the main display was of a French Champagne, he encamped at the Gatwick event and pronounced he wouldn't budge until English wines were included. And, of course, the press was there in no time. They often were when Bob was involved.

It isn't surprising that Camel Valley has often been the choice of wine at both royal and prime ministerial events, with Tony Blair starting the trend. It wasn't a surprise to find Camel Valley Brut being served on British Airways First Class. It wasn't a surprise that they were the first English wine producer to export to Japan, and one of the first to make waves in the USA. And so it was almost a given when, in 2018, Camel Valley became the first English winery to be awarded a Royal Warrant, as a Purveyor of English Sparkling Wines, by Prince Charles, who as well as being the Prince of Wales is also the Duke of Cornwall. And if you had to pick one person to win the International Wine Challenge Lifetime Achievement Award in 2018 – Bob Lindo was that person.

Camel Valley could be said to have truly muscled in on the top table of fizz when their Pinot Noir Rosé won the Best Rosé in the World title at the 2010 World Sparkling Wine Championship. But if this all sounds a bit bling, it isn't. Son Sam now exerts as important an influence as Bob does on English wine. But the family is still rooted in Cornwall. Annie Lindo still prunes every vine on the original planting, as she has now done for 30 years. She reckons to have made 3 million pruning cuts in this time. And blending their Big Time savvy with mud-on-your-boots native wit, they've managed to create English wine's first single-vineyard Protected Designation of Origin (PDO) for their Darnibole Bacchus from their tiny Darnibole parcel of vines at the farm. Is it that special? No. It's a very good Bacchus, but actually no better than their general release Bacchus. But since the whole of Europe hardly has any single-vineyard PDOs, it was very smart of the Lindos to get there first.

And now they've got that, they can get back to what is at the heart of their operation – a highly successful tourist destination in a very popular tourist area – with wines like their Camel Valley Rosé scented with strawberries, their Pinot Noir Rosé Brut, always a little softer than the opposition with soothing foam, and Bacchus whites that really do bristle with the freshness of elderflower and the prickly scent of a West of England hedgerow.

Three Choirs Vineyards

Ledbury Road, Newent, Gloucestershire GL18 1LS
www.three-choirs-vineyards.co.uk Telephone 01531 890555
First planted 1973; vineyard area 28 hectares (70 acres);
total average annual production: approx. 200,000 bottles
At the winery: daily tours, private tours, group bookings, tastings,
brasserie, seasonal events, accommodation
Sales: cellar door, online, local and national stockists
Oz recommends Classic Cuvée NV✳, Siegerrebe, Pinot Noir Précoce
and Ravens Hill Red Blend

Three Choirs in Gloucestershire doesn't often hog the headlines. The winery has been quietly making significant amounts of England's most 'easy to enjoy' wines for 30 years now, always with lots of fruity flavours, never over-priced, English happy juice.

Yet there was one time when they grabbed the publicity bull by the horns. In the 1980s and early '90s Beaujolais Nouveau Day in November used to be one of the biggest days in the English wine calendar – not because people drank any English wine, but because we drank bucketfuls of the brand new, eight-week old red wine from France's Beaujolais region. So, Three Choirs at their patriotic best would release an English 'Nouveau', a 'New Release' on the same day as the French brouhaha hit town. It was white, not red, but it was a lovely drop, and hopefully inserted a small pinprick into the bombastic Beaujolais bandwagon. And there was another time. When I was filming Oz and James Drink to Britain with my hydrocarbon-challenged friend James May, he tasted the Three Choirs Siegerrebe and thought it might be the best wine he'd ever tasted. He bought a case of it. I don't usually hang on every tasting note by James, but I thought the wine was pretty good, too – rich, honeyed, spicy, peachy – and very English.

That's what Three Choirs have always been very good at doing – making wines that are very English. While most other vineyards are falling over themselves to plant trendy varieties like Chardonnay and Pinot Noir, Three Choirs hang on to grape varieties like Siegerrebe and Madeleine Angevine, Rondo and Regent, and even plant up new spicy, fruity varieties like Orion and Solaris. 'We have to keep our style', they say, and if the rest of England is racing off toward drier styles, Three Choirs is quite content to move much more slowly. There are good hard-nosed arguments for it, too. Martin Fowke, the winemaker at Three Choirs for the last 30 years, says that ripening most grapes isn't a problem on the warm red sandy loam soils they have. He says they're on the edge of the famously fruit-friendly Vale of Evesham which is protected from adverse influences on three sides by the Cotswolds, the Malvern Hills and the Welsh mountains, and that the sandier soils anyway give fruit character at the expense of steely backbone. Three Choirs started off as an apple and blackberry farm before turning to grapes in the 1970s, and its wines are still famously fruit-led in flavour. And they sell for very fair prices, without a massive margin for loads of new investment.

Touring through the vineyards with Martin – there a lot of old vines in the ground. They have almost 30 hectares (75 acres) of vineyards and Martin reckons he can afford to replant about 1.2 hectares (3 acres) a year without upsetting the supply of grapes to the winery. After all, if you rip out the vines and replant the land, it might be five years before you get a good full crop again. Even doing 1.2 hectares (3 acres) at a time means he may be missing the grapes from up to 6 hectares (15 acres) each vintage. Add to that the fact that the old vines need re-trellising, and that's he's in transition to sustainable viticulture, removing all herbicides for a start, and you can see why this successful winery doesn't want to change direction too abruptly.

And anyway, Martin feels that the English style of wine that he makes, with no attempt to ape other nations like France, is perfect for now. 'English wine is right for our times.' A generation ago, no one would even eat an English Cox's apple because of the French Golden Delicious – Le Crunch – and everyone wanted oak and alcohol in their red and white wines. English wine was seen as thin and weak. Now low in alcohol, no oak, fresh fruit-led, low carbon miles – that's what consumers want. 'We've spent a long time getting our English identity.' Don't expect a change any time soon.

But evolution, yes. Luckily Three Choirs has good plantings of Bacchus and is putting in more. The quintessential English grape. Perfect for them. And they take English red wine surprisingly seriously. Martin feels that the English red wine revolution got side-swiped by the Sparkling Revolution. But Three Choirs has kept at it. He makes delightful, scented light red Pinot Précoce, but his real talent is in coaxing a thing of delight and beauty out of Rondo and Regent. Rondo can be a dark, tough old thing but you need to keep your testosterone in check making Rondo wine. It has lots of colour and dark sloes and damson fruit so treat it gently. Blend it carefully with Regent and enter it for the 2019 WineGB Awards – where it might win the very English trophy for Best Red Wine Blend, and be just as good a drink as the much vaunted Pinot Noirs.

MORE VINEYARDS AND WINERIES

Lyme Bay Winery

Shute, Axminster, Devon EX13 7PW
www.lymebaywinery.co.uk Telephone 01297 551355
Total average annual production: 200,000 bottles
At the winery: Tastings, events, visitor centre open 2020
Sales: cellar door (English wine, fruit wine, cider, mead, rum and gin), online, local and national stockists, export
Oz recommends Classic Cuvée✳ and Shoreline White

The first Lyme Bay wine I had wasn't made from grapes at all. I had been making a film about learning how to swim at Seaton in

Devon, and half frozen to death, I gratefully gulped down a sweet concoction made from whatever fruits had momentarily come to the winemaker's hand. The most recent bottle of Lyme Bay I had was gratefully gulped down in Bournemouth in a howling gale after I had completed an exhausting yet exhilarating 250-mile charity bike ride. It was an extremely sophisticated traditional method sparkling wine that absolutely hit the spot. Lyme Bay still makes non-grape wines, as well as meads and ciders, liqueurs, gins and pretty much anything you can pour down your throat. But the still wines from quality grapes – they make a good Bacchus – and the sparkling wines are what impress me the most.

Polgoon

Polgoon Vineyard, Rosehill, Penzance TR20 8TE
www.polgoon.com Telephone 01736 333946
First planted 2003; vineyard area 5.6 hectares (14 acres);
total average annual production: 25–30,000 bottles
At the winery: daily tours in summer, small tours in winter, tastings
all year in the shop, café June–Sept for lunch, cottage on site
Sales: cellar door, online, local and national stockists, export
Oz recommends Bacchus and Sauvignon Blanc

Sauvignon Blanc isn't exactly mainstream in the UK yet, and it's fascinating how the examples I like best often come from wineries I hardly know. Well, you can't get much less mainstream than Polgoon. Just outside Penzance on the road to Land's End, it's the kind of locality where you say don't tell me about the soil but tell me does it ever stop drizzling and misting over long enough to ripen wine grapes? Polgoon does have a fair bit of shelter from the worst of the weather. In fact, it has a lot more than 'a bit of shelter'. A significant part of the vineyard is under polytunnels.

Suddenly it makes sense. Protected from any adverse weather, Polgoon's polytunnel vines bud, flower and ripen far more quickly than those on the outside. And that Sauvignon which would never ripen in the teeth of a Cornish gale is snappy and fresh and delicious coming from the protective warmth of the polytunnels.

Sharpham

Sharpham Wine Ltd, Ashprington, Totnes, Devon TQ9 7UT
www.sharpham.com Telephone 01803 732203
First planted 1981; vineyard area 15 hectares (35 acres);
total average annual production: 60,000 bottles
At the winery: tours (guided options need advance booking), tastings,
Sharpham cheese, The Cellar Door Kitchen
Sales: cellar door, online, local and national stockists, export
Oz recommends Madeleine Angevine

It almost wouldn't matter what the wine they make at Sharpham is like. The site of the vineyard on a great loop in the River Dart is so thrillingly beautiful that cold tea would taste delicious sipped on the banks of the river, benignly overlooked by a great Palladian mansion on the top of the hill. The land slopes down to the river like a broad shoulder pressing the water toward the deep, dark woods on the far bank. But only half the slope is covered in vines – where the land is angled toward the evening sun.

It looks as though this might be the perfect protected spot for Pinot Noir and Chardonnay but this is Devon, not Kent, and it's Madeleine Angevine which works best – there are four different versions, including one released at just a few weeks old as a tongue in cheek challenge to France's Beaujolais Nouveau. There are also some of the Germanic varieties (like Bacchus) and some Chardonnay and Pinot Noir is bought in from a neighbouring vineyard. Most of the wines are light and refreshing, fabulous for picnics by the river eating Sharpham's excellent cheese. But bizarrely Sharpham also make some dark, powerful chewy Cabernet Sauvignon and Merlot, from grapes grown under polytunnels at nearby Beenleigh Manor.

Facing page above: Running a vineyard can be lonely work. Annie Lindo has pruned every one of these vines at Camel Valley for the last 30 years.
Below: The Sharpham vineyard is blessed with a stunning location on a broad south-west-facing oxbow bend of the River Dart.

THAMES VALLEY AND THE CHILTERNS

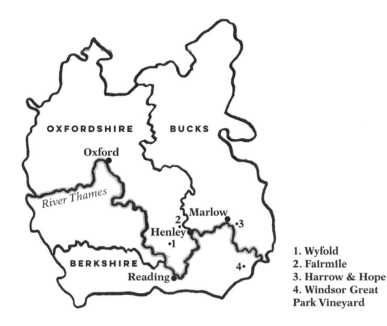

1. Wyfold
2. Fairmile
3. Harrow & Hope
4. Windsor Great Park Vineyard

Since this region spreads west to Hungerford in West Berkshire (where Alder Ridge make excellent sparklers) to Chipping Norton in north Oxfordshire, and east to Milton Keynes in Buckinghamshire, it can be seen as a bit of a hotch-potch. The sites that would seem to be holding the greatest potential for vines are on the slopes of the Thames Valley itself, and on the gravel terraces left behind over thousands of years actually in the river valley. In Oxfordshire and Buckinghamshire the River Thames even carves its way through the edge of the Chiltern Hills. These hills are the same chalk that is proving such a hit with vineyards in Kent and Hampshire. But the Thames Valley has a big obstacle. People like living there. Towns, villages and hamlets crowd the river banks, and vacant land available for vineyards is rare. The slope above Marlow which Harrow & Hope occupy is suitable for vines across hundreds of acres but the trouble is that most of those acres are people's back gardens. But vineyards are being developed. Fairmile and the very new Hundred Hills near Henley-on-Thames show that even where land is expensive, people are prepared to invest. I've noted various interesting chalky or gravelly sites along the Thames near Pangbourne and Goring-on-Thames. And Her Majesty the Queen knows a good investment when she sees one. She now has a vineyard of her own in Windsor Great Park.

Fairmile Vineyard

Fairmile, Henley-on-Thames, Oxfordshire RG9 2LA
www.fairmilevineyard.co.uk Telephone 01491 598588
First planted 2013; vineyard area 2.98 hectares (7 1/3 acres);
total average annual production: approx. 15,000 bottles
At the vineyard: visits by appt, public open days, glamping
Sales: cellar door, online, local and national stockists
Oz recommends Classic Cuvée✴ and Rosé✴

I've known the vineyard site ever since I was an undergraduate. The Fairmile is a dead straight track of main road racing down into Henley-on-Thames and I've been glimpsing these steep, tempting slopes to my left every time I drove from Oxford to London. And in the last dozen or so years, I've got to thinking, surely they would make a fantastic vineyard. Too steep, some experts told me, too expensive inside the Henley town boundaries, others said. So I kept glancing across at these inviting bare grass fields.

Then, in 2011, I thought, hang on, something's up. They're planting vines. And they were. Jan Mirkowski was a telecoms guy from Marlow, just down the Thames Valley. He wanted to get out of telecoms when his daughter was born in 2010. Since he was already a home winemaker, he thought he would have a go at planting a vineyard. Not easy in Marlow. No one wants to sell you their land (the only plot that has come onto the market was snapped up by Henry Laithwaite for Harrow & Hope). And then, Melbury House, on Fairmile, came up for sale and it included 5.6 hectares (14 acres) of that fabulous slope. Jan bought it, and, good telecoms guy that he is, attacked this west to south-west-facing, 1 in 3 slope, with a German planting machine that could plant 1000 vines an hour with the help of 13 satellites in the European, American and Russian satellite systems – speedy and accurate to 15mm. If only the military guys knew!

I walked up the vineyard – that's no joke at 1 in 3 – from the perfectly maintained flint wall at the bottom of the slope, and stubbed my toe on a fair few flints on the way. The bright sun is burning down directly onto the vines. Jan says they get the early sun, the midday sun and the evening sun, because the Stonor Valley dips away to the north-west. And that steep slope may seem a sweaty schlep too far in high summer, but come September and October, the slopes receive 30 per cent more direct sunlight than a flat vineyard does. And that's the difference between grapes that are not quite ripe and just spot on. And certainly it showed with Jan's first amazingly serious, lean but intriguing Classic Cuvee based on the pretty cool 2015 vintage. which won Decanter magazine's English sparkling tasting in 2019. And it showed in the soft scented entirely delightful Rosé, based on 2015 again. With a beautifully pale green 'Henley' presentation, crossing a bottle and an oar for a logo, I asked Jan whether anyone else was going to develop the equally enticing slopes closer to Henley town. It seems someone is. He is.

The Laithwaites

Laithwaite's Wine UK headquarters: One Waterside Drive, Arlington Business Centre, Theale, Reading, Berkshire RG7 4PL
www.laithwaites.co.uk Telephone 03330 148200
Sales: English wines sold since 1975 via Laithwaite's Wine and The Sunday Times Wine Club

---

Are the Laithwaites Britain's most successful wine family? Probably. Tony Laithwaite and his wife, Barbara, have built a remarkable empire of different wine-selling operations. It's supposed to be very difficult to make any money in the wine business. Well, yes, it probably is, if you just stick to the traditional ways of selling wine by the bottle. But Tony's mix of selling direct from the vineyard and forming clubs full of devoted members has changed all that. His brilliance has always been to involve the customer in the whole experience of wine, making them feel a part of the process, from planting a vineyard right through to pulling the cork and sharing the wine.

They talk a lot about family at Laithwaite's, but they don't just mean the Laithwaites themselves – it's everybody who works for the various companies they control, and everybody who joins their clubs – most famously *The Sunday Times* Wine Club – and buys wine from them. It's very fashionable now to talk about selling an 'experience', not just a bottle or a glass. That's what Laithwaite's has been doing since the first day Tony set up in a cobwebby railway arch in Windsor, just down the road from the Castle.

But Laithwaite's is important in another way too. Right from the early days, way back in the 1970s, the company championed English wine and listed any decent ones they could find – not easy, but wines like Adgestone from the Isle of Wight and Spots Farm from Tenterden in Kent were early beneficiaries of a Laithwaite's listing. Even so, Tony was always convinced that southern England was poised to make its mark with sparkling wines using the grape varieties from Champagne. He was quite possibly the first person to order sparkling wines from Ridgeview and for a number of years their own label Ridgeview blend, called South Ridge, was Ridgeview's biggest selling wine.

Well, now they've taken a step further – in fact, four steps further. They are now vineyard owners, sparkling winemakers in their own right, and as ever with this crew, innovators, dreamers and 'darers'. They have developed four different vineyards, none of them in areas that were already on a roll. Look forward a generation and the Laithwaite's will be as important for growing and making English sparkling wine as they are for selling the stuff.

Theale

It's not quite fair to say that Wyfold was the Laithwaite family's first excursion into vineyard ownership. Barbara Laithwaite planted Wyfold in 2003, but there had been a Laithwaite vineyard at Theale in Berkshire since 1998. Funny. I was at Theale this year and I don't remember seeing a vineyard. Well, there isn't one any longer. There's a car park instead. And before that there was a pile of building waste and rubble and bricks from when the Laithwaites built their office down by the station. And this is such a Tony Laithwaite 'mad as a box of frogs' idea. He made the builders compress the building waste into a south-facing slope, planted 700 Chardonnay vines and then required the Laithwaite employees to adopt them, care for them and love them as their own. And then he made the resulting juice into one of England's best sparkling wines.

So much for the French idea of terroir. Obviously you can make terroir wherever you want where the sun shines. So long as you can locate a few bricks and lumps of mud and some tiles and old electric wiring and ensure there's a main railway line a few yards away with trains whizzing by to stop any danger of frost settling. Honestly! I had a few bottles. They were fabulous. When Laithwaite's moved their offices in 2016, the new owners wanted a car park, not a vineyard. So the vines were sold to an outfit in Devon. I hope that Tony told them to mix a few broken bricks into their terroir.

Wyfold

Vineyard: Wyfold Lane, Wyfold, Oxfordshire RG4 9HU
Business address: The Chalet, Peppard Common, Henley on Thames,
Oxfordshire RG9 5EH
www.wyfoldvineyard.com Telephone Barbara Laithwaite
07768 652636
First planted 2003: vineyard area: 2.2 hectares (5½ acres);
total average annual production: approx. 15,000 bottles
At the vineyard: visits by appt
Sales: via laithwaite's.co.uk and Barbara Laithwaite
barbara.laithwaite@directwines.com
Oz recommends Brut✻ and Brut Rosé✻

Wyfold is different to the vineyard at Theale. It's a real field and it's here for the long haul. And if Theale was a joke that turned into an unlikely but whopping success, Wyfold was a hobby that became a serious contender for making one of England's most impressive, most memorable sparkling wines. They say that the best wines come from grapes that have to struggle. Well, they struggle at Wyfold. If the depths of France are called *la France profonde*, this is *England profonde*. The country lanes are tiny and twisting, the trees hem you in so closely you pray a car won't be coming the other way. And the two closest villages to Wyfold both have a village green and their own cricket pitch. That is English.

And Wyfold is high and it's stony. In fact, it's so stony that when they tried to plant vines, diggers and spades couldn't penetrate the soil. They had to use an old-fashioned auger – and that's a really old-fashioned Archimedes screw kind of an implement. No consultant would ever recommend starting a vineyard on soil that difficult. But Barbara didn't have a consultant. She had a friend who was widowed and who had always wanted to start a vineyard with her beloved, and so Barbara sort of stepped in and said 'OK, let's have a go'.

The land was, obviously, next door to her friend's home. It was ridiculously stony. It was way higher and colder than any consultant would condone. But Barbara planted a hectare with 4000 vines and quickly learned how cold it was. She ended up pruning all the vines herself. Every two hours she needed to thaw out as it was so cold. And when the vines gradually took root, she would be picking her grapes later than any other vineyard in southern England – 'sometimes we just run out of summer.'

Yes, summer's normally long gone by November, and the grapes are sometimes picked as late as that. But, wow, it's worth it. The classic white fizz and the rosé have a haughty beauty about them which is awe-inspiring rather than forbidding, because there's so much flavour and character coiled tight, but certain to release and relax. Barbara said that all the wines, even in easy years, have a steel rod running through them. But it's more beautiful, more glinting than steel, sharper and more thrilling than a mere rod. When England took on New Zealand in a Sparkling Wine World Cup in 2015 to add a bit of fizz to the cricket, Wyfold won because it was so supremely, so unmistakeably, proudly English.

Harrow & Hope

Marlow Winery, Pump Lane North, Marlow,
Buckinghamshire SL7 3RD
www.harrowandhope.com Telephone 01628 481 091
First planted: 2010; vineyard area: 6.5 hectares (16 acres);
total average annual production: approx. 40,000 bottles
At the winery: tours and tastings by appt through the website
Sales: cellar door, online, local and national stockists, export
Oz recommends Brut Reserve✳ and Blanc de Noirs✳

This is the beating heart – or should I say, 'thumping heart' of their winemaking adventure. Henry Laithwaite (Tony and Barbara's son) and his wife Kaye run it, very hands on. Harrow & Hope is nestled just above the town of Marlow in the Thames Valley, and, to be honest, Henry didn't expect to end up here. He was making wine in Bordeaux and thought his future lay there. But once their children arrived he and Kaye began to pine for England again. His mother had started making remarkable fizz from Wyfold, near Henley-on-Thames. Could he come home and be a winemaker, too? Well, there's not much land available to buy in a smart commuter and foodie town like Marlow. But there was a bit of farmland on offer which had been used for exercising horses. It was on a south-facing slope, not too high above a bend on the River Thames, the subsoil was solid chalk, and the topsoil liberally dosed with gravel and flint. Wonderful conditions to grow grapes, but tough to cultivate.

The winery's name, Harrow & Hope reflects that boney, challenging soil – you harrow the land and just hope the flints don't break your machinery – sometimes they do. And the wine's neck capsule bluntly states the Laithwaite philosophy, 'Hard Work. Good Fortune'. But that seems to be what this committed pair are after. Being a member of the Laithwaite clan means you might be tempted to take an easy route to market. But they're not. They don't sell most of their wine through the various family operations. They aren't quite organic but are as sustainable as they can be. They've never used herbicides. That makes it pretty challenging to develop a vineyard from scratch, and control grass and weeds.

Henry relates it to his family life. 'If I live here, would I let my kids run around with pesticides on the vines? No, I wouldn't. But if my child was ill, would I take him to the hospital and say – do whatever you have to? Yes, I would.' So you are as sustainable as you can be, and edge toward being organic. Even biodynamic. That's attractive, because as Henry says, even with organic you are trying to kill things whereas with biodynamics you are just trying to out-compete.

And does he interfere a lot in the winemaking? Well, he says, with red wine, the vineyard site is everything; once you get the grapes in it's quite simple. Making sparkling wine is much more technical.

But you can see he's tempted. His wines already exhibit a thrilling vibrancy from this flint and chalk-fed fruit. His imaginative winemaking adds real class and style, right from the very first crop in 2013. But Henry says 'If you can make great wine with more natural methods it gives you so much more satisfaction.' And if this beautiful patch of land perched above the Thames is where he's decided to make his family's home, well, less technique and more nature may be where he's headed in the long term. In 2019 he decided to only use the vineyard's native yeasts for the wine he ferments in barrel. That's a start.

Windsor Great Park Vineyard

Windsor Great Park, Windsor, Berkshire
www.windsorgreatparkvineyard.com
First planted: 2011; vineyard area: 3 hectares (7½ acres);
total average annual production: approx. 14,000 bottles
At the vineyard: by appt for harvest
Sales: via laithwaites.co.uk (online and retail), export (via
production@directwines.com)
Oz recommends Windsor Great Park Vineyard✶

You can't deny that this is a special vineyard. It's not that the site is particularly perfectly suited for growing vines, although it is on a gentle, south-facing slope, and there is a tranquil, soothing lake at the bottom of the vines, the Great Meadow Pond, which can only help to temper frosts and encourage the Pinot and Chardonnay grapes to ripen. The soil isn't special and a few bits of the slope could be described as somewhat cloddish, but there's a fair bit of sand mixed in with the clay, and despite chalk being the soil everyone goes gooey-eyed about, there's a lot to be said for sandy clays on a gentle, south-facing slope.

But this all misses the point. We are in Windsor Great Park here, just down the gallops from Windsor Castle. And not just any old bit of the Park, where you might run across the odd member of the public like you or me. This is the private bit, the really private bit. When I arrived at the gate to visit, a police car drove in front of me, and the gates shut again, as I gazed up at the security cameras. The gate then opened and I drove through the fields and meadows, past the relatively private cricket pitch and golf course, and there was the police car again, beckoning me to turn onto a smaller track, past signs which said in no uncertain terms – this is the private part.

There would never have been a vineyard here if Laithwaite's hadn't messed up a wine order to Windsor Castle. Anne Linder, one of Laithwaites's top managers, went to apologise to the courtier involved and bought him a bottle of Laithwaite's Theale sparkling wine. So he gratefully took her on a tour of the ramparts. As she looked down at one particular steep, south-facing slope below the walls, she said, 'Ooh, that would make a lovely vineyard'. And thought no more about it. Then three months later, the courtier rang Anne and asked if she was serious about her vineyard idea. 'Well, … yes.' 'Excellent. Prepare a presentation for this Friday.' This put Tony Laithwaite into overdrive. They presented a plan on Friday and by the following week the Castle rang again to give the green light.

They were offered five sites, but this one, in the private far corner, made the most sense. It really was private – it could barely be seen except from the lake. It was next door to the old kitchen gardens in which a greenhouse called the Vinery had a massive Black Hamburg vine whose record harvest was over 1000 kilograms of grapes – from a single vine. And at the top of the slope there was a derelict old building called the Boathouse. Just crying out for some loving restoration.

That will come soon, but the vines are up and running, planted in 2011, and producing an average of about 14,000 bottles of sparkling wine each year – though 2018's bumper harvest resulted in 29,000 bottles. The label is a deep, classy Oxford Blue and three shades of gold. Tony wasn't sure what to call the vineyard until the Castle said 'Presumably you'll call it Windsor Great Park Vineyard'. 'Well, of course, we will.' Tony started his wine career in a railway arch at Windsor, so he's bursting with home-grown pride. But he's not the only one. Her Majesty has started serving Windsor Great Park at receptions and has been overheard telling Heads of State 'We have our vineyard, too.'

EAST ANGLIA

1. New Hall
2. Giffords Hall
3. Flint
4. Winbirri

NORFOLK   Norwich

•4

•3

CAMBS

SUFFOLK

Cambridge

•2

Ipswich

BEDS

ESSEX

HERTS

Chelmsford• 1

If there's one area of Britain that has quietly been smooching along under the radar, it has to be East Anglia. Winery owners as far away as Kent, Devon and Cornwall have known about the quality of the East Anglian grapes for a generation or more. But few people outside the wine business knew that some of Britain's best Pinot Noir and Bacchus grapes were coming from such unlikely places as the Essex coast. Well, the cover has been broken.

Major vineyard owners like New Hall are now beginning to use more of their excellent grapes to make into their own wine. Flint Vineyard on the Norfolk–Suffolk border have made some award-winning Bacchus as well as an impressive sparkling wine using tank rather than bottle fermentation to create the bubbles. And Winbirri, near Norwich in north-east Norfolk, swears that Norfolk is the best place in Britain to grow Bacchus and it carried off one of the Decanter World Wine Awards top trophies to prove its point in 2017.

Weather is the chief element that southern East Anglia has in its favour. Despite the Essex soils frequently being heavy and thick with clay, the rainfall is low. And despite temperatures rarely reaching the levels of Kent and Sussex, sunshine hours are long.

Interestingly, the soils often become more suitable as you move north through Suffolk into Norfolk, but the weather does become more challenging, with greater influence from the North Sea, and less protection. Some of Britain's best maritime barley is grown in north Norfolk. But quality will out, and the new wave of Norfolk vineyards and wineries seem to be aware of the challenges. And remember – in the very first International Wine Challenge in 1983, the winning wine was a Rivaner (Müller-Thurgau) from Pulham St Mary in Norfolk.

New Hall Vineyards

Chelmsford Road, Purleigh, Essex CM3 6PN
www.newhallwines.co.uk Telephone 01621 828343
First planted 1969; vineyard area 43 hectares (106 acres);
total average annual production: approx. 130,000 bottles
At the winery: visits Mon– Sat, groups by appt, guided groups and
open tours, events such as Spring wine launch, Sept food and wine
festival and Christmas Wine Sale, weddings
Sales: cellar door, online, local and national stockists
Oz recommends Rosé Brut✱, Signature, Bacchus Reserve and Pinot Gris

---

I'm standing in the middle of the postcode most intensely planted with vines in the whole of England. CM3 6. It doesn't sound that encouraging. Where is it? OK. I'll try again. I've been driving east from Chelmsford in Essex, and just past the village of Danbury, the air changes, the sky changes. The air suddenly seems to smell fresher, a faint tang of ozone and waves breaking on the shore. And the sky. How can a sky become bigger, but this sky seems to: it seems broader. It seems bluer. It … well, it seems bigger.

Now I'm in the churchyard at the top of Purleigh Hill gazing out to the bullrushes and the white sails of boats bobbing on the River Blackwater. And now I see them – vines to the east, north and west. I follow the lane round past the church, and despite the unpromising sounding village of Cold Norton being signposted – out toward the east and toward the marshes of the River Crouch in the south – more vineyards. And more every year, as this part of Essex sets out to prove it may be the best area of all Britain for vineyards. The vines I see from Purleigh facing northward are mostly those of New Hall. This really is the granddaddy of English commercial vineyards, and the quality of its grapes must be one of the main reasons the Crouch Valley area is now teeming with wine activity. And if you say, hmm, I've never tried a wine from Essex, well, actually, you probably have. If you've enjoyed mouthwatering, hedgerow-fresh, elderflower-scented Bacchus wine from wineries in most of the counties in southern England, there's a good chance that some or all of the grapes came from New Hall in Essex. If you've been thrilled by the quality of a lot of the pink sparkling wine offered by numerous wineries to the south and west of London, it was probably New Hall Pinot Noir which was giving the wine its scent and flavour – and its ripeness.

Ah, yes. Ripeness. You may not immediately think that the Essex coast, jutting out into the chilly North Sea, is going to be perfect vineyard land. Will it be hot enough? If you've been listening to almost everybody about how important it is for English vineyards to be situated either on chalk or on warm greensands and sandy clays, the prospect of a vineyard that is pretty much solid claggy London clay may not appeal. But we haven't mentioned the most important factor. It's dry.

East Anglia, especially on the Essex and south Suffolk coast, is the driest place in England. Counties like Kent may get more sunshine hours, but it's the lack of rain that is crucial. Southern England has generally been warm enough to ripen grapes all through the 21st century. It's frost and rain that will usually wreck a harvest. New Hall and its Crouch Valley neighbours are low-lying and near the sea. Frost hardly ever forms here. In April 2017 many vineyards in Britain lost 50 per cent or more of their grapes through frost. New Hall only lost 5 per cent, and that was only in the vines along the hedgerows at the edges of the fields.

And that rain? Well, Essex gets rain alright. But where the vines grow, especially between the Blackwater and the Crouch, they often only receive 500 millimetres (19½ inches) of rain a year – that's low. It can be raining in Maldon and Chelmsford, only a few miles away, but not at New Hall. The rain follows the estuary, and the typical south-westerly rain clouds general miss this south-east corner of England. A neighbouring vineyard owner has even installed irrigation because his vines get so dry in years like 2018 that they become drought-stressed. And as for clay, well, it is at its worst when it's wet and cold. Clay, when it's warm and dry, is proving to be a decent soil in England for such varieties as Pinot Noir and Bacchus – precisely the two varieties that made New Hall famous.

The vineyards at Purleigh have been famous before – King John used their wine to celebrate the signing of Magna Carta in 1215. But in modern times a visit to New Hall lets you experience almost the whole of modern English wine in one place, and under the ownership of one family. The vineyard was planted in 1969 by Bill and Sheila Greenwood from a job lot of Reichensteiner vines they had bought at a farm auction for 23 pence a vine. Their first vintage was 1971 – 18 bottles, made and bottled by Sheila in her kitchen. By 1973 they were one of the first vineyards to plant Pinot Noir and Pinot Gris, and in 1976, they were pioneers with Bacchus, too. In 1985 one of Britain's first traditional method sparkling wines was made at Lamberhurst from New Hall grapes. And so it goes on.

And these old vineyards are still there. Almost all of the original English commercial vineyards have been pulled up, but not at New Hall. The old Bacchus vines look wonderfully bedraggled in a chaotic weed-clogged field, but Andy, the devoted vineyard manager, says they're too precious an inheritance to let die and he's reviving them with an ingenious method of trunk replacement. And after that's done, he'll deal with the daisies and dandelions. And we won't be tasting quite so much New Hall fruit under other wineries' colours any more. New Hall are making more and more of their own wine. Expect to see a 'Proud to be Essex' movement on the march sometime soon.

MORE VINEYARDS AND WINERIES

Flint Vineyard

Middle Road, Earsham, Norfolk NR35 2AH
www.flintvineyard.com Telephone 01986 893942
First planted 2016; vineyard area 6 hectares (15 acres);
total average annual production: 30,000 bottles
At the winery: regular prebooked tours and tastings, lunches, events
Sales: cellar door, online, local and national stockists
Oz recommends Charmat Rosé✳ and Bacchus

It was my friend Matthew Jukes, an avid supporter of English and Welsh wine, who prodded me into taking more notice of Flint. They had just released a sparkling wine made by the tank or Charmat method, rather than the more expensive traditional (Champagne) method, which creates bubbles using a second fermentation inside the final bottle. Charmat wines get their bubbles from inducing a second fermentation in the tank, before bottling, and this is a cheaper, and, usually, inferior method of making fizz. But as Ben Witchell, Flint's owner, says, this method preserves the aromas and fruit flavours of the wine, so is entirely suitable for aromatic grape varieties like Bacchus and Solaris. And he's right. The wine is fresh, foaming and delicious, and shows that there may be a future in Britain for this cheaper form of bubbles. Even so, it's the zesty, snappy still Bacchus I like best – partly from grapes grown on their site by the River Waveney in south Norfolk, and partly from Essex fruit, proving, once again, that East Anglia may well be the best place in Britain to grow Bacchus.

Giffords Hall Vineyard

Hartest, Bury St Edmunds, Suffolk IP29 4EX
www.giffordshall.co.uk Telephone 01284 830799
First planted 1986; vineyard area 5 hectares (13 acres);
total average annual production: 30,000 bottles
At the winery: tours open April–Sept, groups and guided
tours by appt, tastings, Al Fresco café, events
Sales: cellar door, including English liqueurs, online, local and
national stockists, export
Oz recommends Classic Cuvée non-vintage✱, Bacchus and Madeleine
Angevine

If the Japanese consumers are looking for something quintessentially English, they couldn't do much better than to buy Giffords Hall Madeleine Angevine from Long Melford in Suffolk. And that's exactly what John and Linda Howard were thinking as they managed to persuade two of Japan's most prestigious department stores to stock their wines. You might expect the ritzier sparkling wine producers of Kent and Sussex to be following this export path, but Giffords Hall is leading the way for UK wine estates. In 2018 40 per cent of their wine was exported. If we are going to have a problem of oversupply of grapes in the 2020s, other producers should be queuing up in Suffolk for lessons in how to export.

And you don't have to be just growing Chardonnay and Pinot Noir. Madeleine Angevine is regularly the first wine I taste at Giffords Hall, and often my favourite. It's adaptable too. When the Howards wanted to make a very pale rosé out of the rather potently coloured Rondo, they blended it with Madeleine Angevine to make a very fair approximation of a pale Provence rosé – and they sell loads of it in magnum bottles. They make red wine, too, I particularly like their St Edmondsbury red, named after the local cathedral diocese, which they manage to export to the Vatican in Rome.

Winbirri Vineyards

Bramerton Road, Surlingham, Norwich, Norfolk NR14 7DE
www.winbirri.com Telephone 01508 538974
First planted 2007; vineyard area 13.5 hectares (33 acres);
total average annual production: 75,000–100,000 bottles
At the winery: group tours and tastings by appt May–Sept,
public harvest day events
Sales: cellar door, online, local and national stockists
Oz recommends Bacchus and Solaris

When you win an award, it's often how you turn it to your advantage that matters most. There is no point in winning a gold medal and not telling anyone about it. Well, you needn't worry about Winbirri not letting the world know about what's happening in deepest Norfolk. In 2017, Winbirri's Bachus 2015 wowed an international panel of judges at the Decanter World Wine Awards and won 'Best Value White Single Varietal' wine against the rest of the world. The English press and news websites went mad for what is certainly a very attractive, bright, elderflower-scented, zesty white. Lee Dyer, Winbirri's owner, says that he received 10,000 emails asking for wine in the first two days after the award was announced.

Winbirri makes red wines too, as well as sparklers, and Dyer thinks that he has a special local climate along the River Yare which particularly favours good September weather, allowing him to ripen Pinot Noir and Chardonnay. But it is the Bacchus that shines here, and Winbirri's trumpeting of their own success put the name Bacchus in front of millions of potential consumers. All England should say thank you.

MIDLANDS AND THE NORTH

1. Halfpenny Green
2. Leventhorpe

This is the biggest wine region in the UK by far, but it is thinly spread with vineyards – which isn't surprising. Vines need warmth and a reasonable amount of dry weather. Counties like Lancashire, Yorkshire, Derbyshire and Cumbria are not best known for their long, balmy days and mild refreshing rain showers. Not yet, they aren't. But we may not have to wait all that long as climate change stalks remorselessly north. And there's a lot of very suitable vineyard land, particularly in Yorkshire and Lincolnshire, which at the moment is often a little too high or too cool. But you only have to taste the transformation that has occurred at Leventhorpe in Leeds this century to believe the limestone and sandstone soils that abound right up to Middlesbrough on the north-east coast may yet come in to play in the non-too-distant future.

Yorkshire already has 20 vineyards, with Leventhorpe, Yorkshire Heart and Little Wold leading the way. And in the meantime, pioneers are teasing out little patches of protected farmland – at Dunesforde Vineyard, near York, they have vines at just 15 metres (50 feet) above sea level just west of the River Ure and supposedly 'in the rain shadow of the Pennines'. Hmm. That's a new one for us southern softies. But they're not the only believers in the Pennine rain shadow. Further down the Ure Valley, Yorkshire Heart has been growing vines and making wine since 2006. They now have almost 6 hectares (15 acres). East Yorkshire could be promising, too. Laurel Vines has found south-facing chalk near the River Hull at only 8 metres (26 feet) and having started off with Ortega, Solaris and Rondo now has Chardonnay and Pinot Noir, sheltered from the East Coast breezes. At South Cave, down toward the Humber, Little Wold are thriving on a protected slope of chalky Yorkshire Wolds soil – and they've also got some Chardonnay and Pinot Noir. There are more sites like these for those brave enough to take the plunge. Even Derbyshire has an award-winning vineyard at Renishaw Hall.

There are many more vineyards in the Midlands, and the weather is warmer, and in Shropshire and Staffordshire there is some evidence of a rain shadow reducing precipitation. Several of the suggested sites for Roman vineyards are in the Midlands at places like Wroxeter in Shropshire and Wollaston in Northamptonshire. The problem in the Midlands is frequently the soil – it's generally heavy, fertile, often dominated by clay. Vines much prefer infertile soils like limestone and sandstone, which restrict growth and allow fewer bunches of fewer grapes to creep to ripeness. Since there are so many other parts of England whose conditions are well suited to vines, I wouldn't expect the Midlands to play more than an occasional role. But Yorkshire, well, 'God's Own County' and all that...

Leventhorpe Vineyard

Newsam Green Road, off Bullerthorpe Lane,
Woodlesford, Leeds LS26 8AF
www.leventhorpevineyard.co.uk Telephone 0113 2889088
First planted 1986; vineyard area 2.2 hectares (5½ acres);
total average annual production: approx. 16,500 bottles
At the winery: open all year, groups by appt
Sales: cellar door, mail order, local and national stockists, export
Oz recommends Madeleine Angevine and Seyval Blanc

'Leventhorpe is a once-tasted always remembered kind of wine.' That's a quote from me on Leventhorpe's publicity material. Well, it's true, Leventhorpe doesn't taste like any other English wine, but when I made that remark it barely tasted of wine. I would be offered it every year, usually as a blind tasting on TV or radio in Leeds, with the bottle wrapped in brown paper, and the camera crew giggling behind their hands. He'll never guess this. But I always did. One tasting note I made said 'Raw, ferrous, stained by Pennine rain, scarred by the sooty crust of industry.' Yorkshire's finest, tasting unerringly of the rusty decay of post-industrial Castleford.

Well, it's still Yorkshire's finest. But in 2005, something remarkable happened. The grapes were ripe, the wine was delicious and every vintage since has been a Yorkshire delight. But always Yorkshire. I suspect George Bowden's middle name is 'Yorkshire'. And when I told him I finally wanted to visit – he said 'Right. Get out of Castleford toward Swillington, head up Bullerthorpe Lane, past Gamblethorpe Farm, if you reach Woodlesford, you've gone too far.' How Yorkshire can you get?

But I did get there, and I did discover what really does seem to be a magic field of vines. George was once a teacher, and regularly used to drive past this field and in winter the snow always melted here first, and when there clouds all around, this field seemed to bask in sunlight. He discovered the field had been mentioned in the Domesday Book, and farmers used to overwinter their pigs there, because it was warm. So he bought the field, planted the vines on the south-facing sandy loam slope down toward the River Aire, and built a winery. 'Some people might call it a shed. Some people might call it a château.' And it works. George has just over 2 hectares (5 acres) of vines and he always gets a crop of grapes. In 2012 most of England's grapes never made it to harvest. His 2012s were plentiful and won him a gold medal. In April 2017 most English vineyards lost half their harvest to frost. At Leventhorpe not a single bud was harmed.

And I've been keeping my bottle of Leventhorpe Madeleine Angevine 2005 all these years. Cherishing it. Truly believing it was something special. And today I opened it. It was a cracker. Full brilliant gold, with a pure, bright boiled lemon pith acidity, fluffy green undergrowth scent, warmed by a splash of pale dry chocolate. And was there just a streak of mineral dust? Well, yes, that's your Castleford terroir, isn't it?

AND ONE MORE WINERY

Halfpenny Green Vineyards

Tom Lane, Bobbington, Staffordshire DY7 5EP
www.halfpennygreenvineyards.co.uk Telephone 01384 221122
First planted 1983; vineyard area 13 hectares (33 acres);
total average annual production: 55,000 bottles
At the winery: private and public tours by appt, wine-tastings with
a self-guided tour by appt, group bookings welcome, wine and gin
tastings, wine shop, tearoom, restaurant, craft centre, accommodation
on site, fishing pools, events including weddings
Sales: cellar door, online, local and national stockists
Oz recommends Tom Hill White

Just outside Wolverhampton, in the heart of the Black Country in the West Midlands, doesn't sound too promising as the address for a vineyard. On the Shropshire border sounds better, and that's where Halfpenny Green is – pretty close to Wolverhampton but you'd never know it. On the Shropshire border is just right, and it's the site of a surprisingly big vineyard – about 12 hectares (30 acres). They make a series of good wines, mostly from the less fashionable varieties like Madeleine Angevine, Solaris and Rondo – I remember we awarded them several medals in the 2017 International Wine Challenge.

But this isn't why they are so important. It's their work as a contract winemaker for the whole of the Midlands and Wales area that is crucial. Global warming has meant that both regions, previously little regarded as any good for vineyards, now have dozens of small growers – big enough to try to sell their wines but too small to have their own winery. Halfpenny Green (along with Three Choirs in Gloucestershire) provides the facilities. They now have 92 different stainless steel vats making wine for more than 60 growers. You may not know the name Halfpenny Green, but if you're drinking wine from the Midlands or Wales, there's an odds on chance it was made here, just outside Wolverhampton.

LONDON

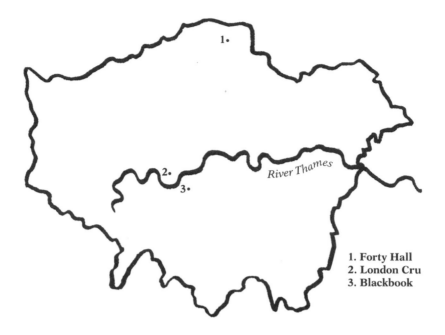

1. Forty Hall
2. London Cru
3. Blackbook

I realise I have tasted quite a few London wines over the years. There was a chap I knew who grew Müller-Thurgau on his allotment near Barnet, someone else made some half decent red from his clutch of vines at Romford and a lady who thought her vines ripened nicely because they were tucked up against the Heathrow perimeter fence.

But the first example I really enjoyed was a Tooting rosé. My brother brought me a bottle and we had it along with a well known Provençal rosé. I thought the Tooting was a much better drink. Chateau Tooting, aka The Urban Wine Company, proudly declares itself as 'making wines from grapes grown in the community'. The company uses whatever quantity of grapes it can persuade owners to deliver from their back gardens in London. Sometimes there will be a serious quantity from an old vine against a warm wall that simply keeps on giving. After all, a single old vine at Hampton Court produces about a tonne a year. But the only proper-sized London vineyard is at Forty Hall in Enfield. As well as Forty Hall, there are several urban wineries based in London which use grapes from further afield. These are ideally placed for Londoners to visit and get experience of a working winery at first hand – even during the lunch hour if you so wish.

One of them, Vagabond Winery in Battersea, is actually attached to the Vagabond Wine Bar and Wine Shop, so you can visit the winery, drink a glass of your favourite wine with some food in the wine bar and then buy another bottle to take home. Seamless. And the wines are good.

Forty Hall

Gardener's Rest, Forty Hall Farm, Forty Hill, Enfield,
London EN2 9HA
www.fortyhallvineyard.com Email: info@fortyhallvineyard.com
First planted 2009; vineyard area 4 hectares (10 acres);
total production: (2018) 9500 bottles
At the vineyard: self-guided tours and tours by appt, tastings in the
vineyard, farm and farm shop open Fri–Sun; monthly farmers'
market; tasting events, team building days for organisations and
corporate business, pop-up food and wine events
Sales: farm shop, online, local and national stockists, export
Oz recommends Brut✳ and Bacchus

I can't be in Enfield. I just can't be. I've paused in a dusty country lane, silent except for the drone of bees, cooled by the overhanging boughs of the trees on this sunniest of summer's day. I've stopped because the odour of the nettles that crowd in on both sides of the track is so powerful I could be back in childhood. And a vole scampers across in front of me.

And I can't be in Enfield. I've walked a few paces more, and through a gap in the hedge ranges a vineyard, acres of it, sloping down toward the line of high old trees guarding it at the bottom. Tranquil, reassuring, calming. Two Red Admiral butterflies dance in the haze by the willow-woven sun shelter. A faint reminder of human activity wafts on the breeze – the whisper of happy chit chat, laughter, as the volunteers move between the vines, pulling off unwanted leaves, raising the trellis wires to help open up the fruit to this ripening sun.

And I am in Enfield – I'm at Forty Hall Vineyard, with the North Circular Road to the south, and the M25 to the north. They say you can hear the drone of traffic if you listen hard, but I listen, and I can't hear it. And I don't want to. As far as I am concerned, I could be in the depths of the Kent or Sussex countryside, far from any main road or town. And that, I realise, is increasingly the point. Forty Hall is not just about growing vines and making wine – although they are the only commercial vineyard within the boundaries of London. Forty Hall is about soothing and healing the human spirit, about mending minds that have been bruised, about boosting and restoring lives that have been scarred.

The vineyard is merely one part of the Forty Hall project – a fully organic farm, which includes orchards, a forest garden, a market garden and rare breed livestock. Its tagline explains its purpose – 'Cultivating a happier and healthier community'. This project isn't here just to make delicious wine – which they do – a fabulous creamy, bright sparkler, and a hedgerow-scented Bacchus are my favourites – it's a social enterprise. Enfield has the twelfth highest rate of depression of any London borough. It has more than 32,000 people living with mental illness. In Enfield the NHS spends £98 million a year on treating depression and anxiety alone.

And so in 2009, an inspirational lady from Hackney called Sarah Vaughan-Roberts decided she would like to establish a vineyard in London. Enfield was the only borough to respond to her request. Her initial 0.4 hectares (1 acre) has grown to 4.04 hectares (10 acres), and is a full-scale social enterprise, eagerly supported by the borough, but also by the community which provides the volunteers who do all the work among the vines. (There's a waiting list of enthusiasts eager to join.) The vineyard has been constituted as a 'not for profit' limited company since 2010 and includes two eco-therapists on its small paid (but part-time) staff, who had just finished working in the vineyard with a group of refugees from Iran and Afghanistan when I visited. Some were in a women's refuge. 'We had so much fun. We can breathe at last here.' I sipped the 2016 sparkler. The bunch of sweet peas arranged on the table before me was so heady and scented I couldn't smell the wine. 'It's the happiest I've felt in my life,' said another volunteer. When you drink Forty Hall wine, you're not just drinking the wine.

TWO URBAN WINERIES

Blackbook Winery

Arch 41, London Stone Business Estate,
Broughton Street, London SW8 3QR
www.blackbookwinery.com Telephone 07816 658471
Total average annual production: 18,000 bottles
At the winery: tours and tastings by appt, open sessions available to
the public, occasional supper clubs, special events including producer
tasting events, spring and summer open days, volunteer list, wine
club, post harvest barrel tasting event for wine club members
Sales: cellar door, online
Oz recommends Tamesis Bacchus, Painter of Light Chardonnay and
Nightjar Pinot Noir

It was my brother who kept urging me to take more notice of
Blackbook Winery. He's a member of Gareth Malone's Battersea
Power Station choir and he kept telling me that there was a
fantastic winery just next door to a charcuterie, and that the choir
were using their wine to wet their whistles – and he's a pretty good
judge of wine.

And then I tasted Blackbook's 2017 Chardonnay. Many winemakers
try to make a Chardonnay that tastes like a top Burgundy but few
succeed. Yet this 2017 had all the toasted hazelnut and savoury
richness of Puligny-Montrachet – and it was made in Battersea
by a young American called Sergio Verrillo from grapes grown
at Clayhill Vineyard on the Crouch Estuary in Essex. Sergio has
worked, among other places, at Calera in California, Ata Rangi in
New Zealand and de Montille in Burgundy – you couldn't choose
better places to learn about Pinot Noir and Chardonnay. Clayhill
is one of the leading growers in Essex, a region increasingly being
touted as England's potentially finest vineyard area. And it all
comes together under a railway arch in Battersea. Urban wineries
have been around in the USA for at least a generation but it's a
new movement in Britain. It makes absolute sense with London's
massive inquisitive market right on the doorstep, and the ever-
increasing desire of Londoners to eat and drink 'local'.

London Cru

21–27 Seagrave Road, London SW6 1RP
www.londoncru.com Telephone: 020 7381 7870
Total average annual production: 10,000 bottles
At the winery: tours, winemaker experiences as part of the Annual
Crush Club, private events
Sales: cellar door (Mon–Fri), online (via www.robersonwine.com)
Oz recommends Baker Street Bacchus and Rosaville Road Rosé

The first time I tried to find London Cru, I did turn into the correct alley, just opposite The Atlas pub in West Brompton, but it looked so unprepossessing that I actually went back to the pub to ask if I had got the right place. Well, I had. The winery and the offices of Roberson, a fine wine merchant, are squeezed into a hidden yard that used to be a gin distillery. But there they were – stainless steel tanks, presses, barrels and all in the heart of London. The wines were excellent, but in those days they were mostly from European mainland grapes, except for a deliciously juicy crisp Bacchus.

Since 2017 London Cru has been an all-English operation, with Alex Hurley, the winemaker, feeling right at home despite having made wines in such disparate places as Burgundy, Barolo and Australia. The focus is on still wine based on grape varieties doing well in the English climate. Alex revels in the possibilities of early-ripening varieties like Bacchus and Pinot Noir Précoce, determined to highlight their character rather than pretend they are something else. He is a real evangelist for English and Welsh wine styles, bemoaning the fact that most wine drinkers still haven't tried an English wine. In 2019 around 25 per cent of the production was a traditional method sparkler, Blanc de Noir from Pinot Meunier.

At London Cru, the door is always open – it's easily accessible from the Tube and Overground – you could pop in after work. And if you like what you taste, you can enroll in the Crush Club and get your hands dirty actually making the wine.

WALES

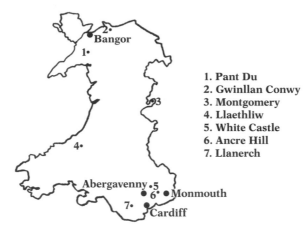

1. Pant Du
2. Gwinllan Conwy
3. Montgomery
4. Llaethliw
5. White Castle
6. Ancre Hill
7. Llanerch

Wales doesn't have many claims to wine fame yet. And the most significant wine act of Wales in the 20th century was to provide the British Isles' last surviving commercial vintage at Castel Coch in the Vale of Glamorgan west of Cardiff. These vines were pulled out in 1920 and until 1952 there was no commercial vineyard in Britain. It's difficult to know whether Castel Coch was really much of a vineyard since most of the wines had masses of sugar added to them and ended up sweet and with a probable alcohol level nearing that of port or brown sherry. *Punch* magazine wasn't impressed and said that it would take four men to drink the wine – the victim, two men to hold him down and one to force the wine down his gullet.

Well, maybe. But that was then. Nowadays the Vale of Glamorgan is looked upon with cautious optimism. Llanerch is a successful vineyard quite close to the original Castel Coch site and the Vale's climate is generally reckoned to have some of the highest average mean temperatures in Britain. Further west, Pembrokeshire is thought of as possibly suitable and there's a bit of activity in Ceredigion, especially near Aberaeron where Llaethliw Vineyard braves frost and hail to make some fair whites. Pant Du in Snowdonia manages both cider and wine and Gwinllan Conwy has a good stab at Solaris white and Rondo red near Llandudno.

But the areas that are generally reckoned to be most suitable are a bit further inland. The Vale of Denbigh, heading due south from the north coast, has got some warm, sandy sites which are attracting attention. Two thousand vines were planted there in May 2019. They've got some impressive sounding Cabernet Noir – well, it's not that impressive – it's a Cabernet-like vine that can ripen in Sweden, so it should be a doddle in Denbigh. The area around Newtown in Montgomery is well protected and is producing some good Bacchus and Solaris whites and pretty fair Rondo reds. Montgomery Vineyard is the award-winning leader here. But it's Monmouth that looks the most promising. Ancre Hill manages to be biodynamic in the Monnow Valley. Robb Merchant, who owns White Castle, says he's looked at 12 new sites for others to plant in the last two years and most of these are in Monmouthshire.

Wales is a small but dynamic part of our national wine movement. When I'm talking about the nation's vineyards and wineries, I do sometimes say 'English and Welsh' wine; but I hope the Welsh won't mind that I mostly use the more concise term 'English' wine. One day we'll be calling it all British wine, but not just yet.

White Castle

Croft Farm, Llanvethrine, Abergavenny, Monmouthshire NP7 8RA
www.whitecastlevineyard.com Telephone 01873 821443
First planted 2009; vineyard area 2.5 hectares (6¼ acres);
total average annual production: approx. 6000 bottles
At the winery: daily tours, large groups by appt, conducted tours
and tastings, lunch platters, events including weddings
Sales: cellar door, online, local and national stockists, export
Oz recommends Brut✳, Pinot Noir Précoce Reserve, Regent and
Rondo

I'm not sure I'd have taken an awful lot of notice of Welsh wine if I hadn't met Robb Merchant, the owner of White Castle vineyards near Abergavenny. Too wet, too windy, too Welsh ... and then in bursts Robb. He's probably the most passionate man in British wine. His belief in Wales and Welsh wine and how it can bring enormous benefit to rural Wales and provide jobs and revenue and security where none currently exists radiates off him as though it were steam blowing out of his ears. He's not the first of the New Wave Welsh winemakers. He's not the biggest. He's not the best known. But Wales are lucky to have him, and the remarkable 21st century revival of Welsh wine has got his thumbprints all over it.

He is not your typical vineyard owner. He was working for Royal Mail, in increasingly stressful roles that he said would have delivered him a big fat heart attack if he'd kept going much longer. And his wife was a district nurse. 'We don't come from a silver spoon upbringing and we value what we have. And I'm a lifestyle millionaire.' They'd managed to buy an old farm with 4.8 hectares (12 acres) of open, north-facing land near Abergavenny in 1993, and if Nicola began harbouring dreams about establishing a vineyard, Robb didn't. But in 2008 a 2-hectare (5-acre) plot just behind the barn was offered for sale. And this site was different. It was quite steep, it was on sandy clay, and it was south-facing. And Robb was getting seriously fed up with the Royal Mail. Within a year Nicola's dream was a gumboots and muddy hands and sore knees reality and within another year, these 2 hectares (5 acres) got the full Robb Merchant attention.

It was his fizz that first caught my attention, in 2014, but it should have been his Regent red, because White Castle's reds, from Regent, Rondo and Pinot Précoce are far fuller of flavour than I'd have expected. In 2019 I drank a Précoce 2012 which for want of a better term, had aged in a positively Burgundian way to a deep, truffley yet red-fruited maturity. An unoaked Rondo 2016, dark and full of bilberry fruit and sappy intensity made me think I would have to re-evaluate my views on Rondo's inherent coarseness. And 2016 was when Robb took on the chairmanship of the Welsh Vineyards Association. This body had actually been founded a month before the English Vineyards Association, in September 1965, but doesn't seem to have done much since. Until now. What Robb realised was that in a small country like Wales, politicians are much easier to grab hold of, and much more likely to realise the benefits to the rural economy that wine could bring. Food and Drink Wales has funding available and the vineyards are getting it.

Robb took the Welsh government minister out into his vineyard as soon as he became chairman. When Robb told her how much he could gross from his small vineyard and how many tourists he could attract to his little winery, and what they might spend, the minister cancelled her afternoon meetings and by the end of the day they'd put a plan together that would provide support for the first four years of establishing a vineyard. The Welsh government would also fund stands at events like the London Wine Fair and the Royal Welsh Show. And Farming Connect – another Welsh government initiative – provided the means for seven Welsh vineyard owners to go on a study trip to the Loire Valley.

Robb didn't waste that Loire trip. He obtained the weather statistics for the red wine area of Chinon and laid them over the stats for his area of Monmouthshire. Climate change showed that between Abergavenny and Monmouth town the weather was now coming into the mid-range of Chinon's statistics. That north-facing field of his, surely it would be too cold? The stats said no. The aspect from east to west said no. The low hills and the copse of trees shielding the site from the south-westerly gales said no. And so in 2019 he planted 1000 Cabernet Franc vines, the same as Chinon's famous red grape. And I won't bet against them producing some bright, crisp crunchy light reds by the end of 2021.

Healthy-looking Phoenix and Rondo vines thrive on the sheltered, south-facing slope above the old barn at White Castle.

MORE VINEYARDS AND WINERIES

Ancre Hill Estates

Rockfield Road, Monmouth NP25 5HS
www.ancrehillestates.co.uk Telephone 01600 714152
First planted 2006; vineyard area 12 hectares (30 acres);
total average annual production: 20,000 bottles
At the winery: guided tours, tastings, prebooked Welsh cheese platter
lunches, self-catering cottage
Sales: cellar door, local and national stockists, export

Gwinllan Conwy Vineyard

Llangwstenin, Llandudno Junction LL31 9JF
www.conwyvineyard.co.uk Telephone 01492 545596
First planted 2012; vineyard area 1.4 hectares (3.5 acres);
total average annual production: 10,000 bottles
At the winery: tours, tastings, special events
Sales: cellar door, online, local and national stockists, export

---

Llaethliw Estate Vineyard

Llaethliw, Neuaddlwyd, Aberaeron, Ceredigion SA48 7RF
website www.llaethliw.co.uk Telephone 01545 571879
First planted 2009; vineyard area 6.9 hectares (17 acres);
total average annual production: 10,000 bottles
At the winery: tours by appt, cottage in Aberaeron
Sales: cellar door, online, local and national stockists

Llanerch

Hensol, Vale of Glamorgan, South Wales CF72 8GG
www.llanerch-vineyard.co.uk Telephone 01443 222716
First planted 1986; vineyard area 2.8 hectares (7 acres);
total average annual production: 12–15,000 bottles
At the winery: daily tours and tastings by appt; restaurant and bistro,
hotel, cookery school, bar, shop, weddings and events
Sales: cellar door

Montgomery Vineyard

Cefn-Y-Coed, Montgomery, Powys SY15 6LR
www.montgomeryvineyard.co.uk Telephone 01686 670301
First planted 2013; vineyard area 1.2 hectares (3 acres);
total average annual production: 9,500 bottles
At the winery: tastings for trade only
Sales: online (via www.discoverdelicious.wales), local and national
stockists, export

Pant Du Vineyard

Ffordd y Sir, Penygroes, Gwynedd LL54 6HE
www.pantdu.co.uk Telephone 01286 881819
First planted 2007; vineyard area 3.6 hectares (9 acres);
total average annual production: 3000 bottles
At the winery: guided tours by appt, group wine tastings (evenings),
café, bar, orchard
Sales: cellar door (including ciders, apple juice, spring water), online,
local and national stockists

RESOURCES

WineGB www.winegb.co.uk
The national association for the UK wine industry. English and Welsh Wine Week takes place in late May and early June and is the perfect time to visit local vineyards and wineries with many events and activities planned around this week.

Regional associations
- WineGB South East/South East Vineyards Association www.seva.uk.com
- WineGB Wessex www.winegb.co.uk
- WineGB West of England www.swva.info
- Thames & Chilterns Vineyards Association www.thameschilternsvineyards.org.uk
- WineGB East Anglia www.eastanglianwines.co.uk
- WineGB Midlands & North www.mercianvineyards.org.uk
- Welsh Vineyards Association www.winetrailwales.co.uk

Other useful addresses
- Plumpton College www.plumpton.ac.uk *Make wine your career. Excellent courses and apprenticeships in winemaking, wine business and viticulture.*
- Sussex Wineries www.sussexwineries.co.uk *Sussex has the largest group of vineyards in England.*
- Vineyards of Hampshire www.vineyardsofhampshire.co.uk *Grouping of vineyards formed in 2013. Annual sparkling wine festival*
- Vineyards of the Surrey Hills www.surreyhillsvineyards.co.uk *A cluster of vineyards located in the beautiful North Downs.*
- Wine Garden of England www.winegardenofengland.co.uk *Association of seven producers and wineries in Kent with a wine trail and visitor experiences.*
- Visit Wales www.visitwales.com/things-do/food-drink/award-winning-wines-wales

Wine touring
- English Wine Tasting & Tours www.EnglishWineTastingTours.co.uk *Tours from central London into the countryside*
- Fizz on Foot www.fizzonfoot.com *Walking and wine tasting tours*
- Grape & Grain Tours www.grapeandgraintours.co.uk *Visits to outstanding wine, beer and gin producers in Surrey and Hampshire*
- Great British Wine Tours www.greatbritishwinetours.co.uk *Exploring the best vineyards, breweries and artisan food producers in Sussex, Kent and Surrey*
- WineCellarDoor www.winecellardoor.co.uk *Who knew that more than 200 vineyards welcome visitors? Want to visit vineyards with vegan wines or with electric car chargers? Information on visiting vineyards and wineries and making the most of your visit.*

Supporters of English wines
- Butlers Wine Cellar www.butlers-winecellar.co.uk
- *Two shops in Brighton. Huge range of English wines, sparkling and still. Mail order and online.*
- The Coral Room www.thecoralroom.co.uk *Wine bar, and more, in the Bloomsbury Hotel in the heart of Bloomsbury, London with a rotating list of English sparkling wines served by the glass.*
- English Wine Centre www.english-wine-centre.co.uk *Located at Alfriston*

on the edge of the South Downs National Park. Under new ownership for 2020 with refurbished facilities. Large selection of English wines to taste and buy. Wine tasting, weddings and events. Luxury accommodation, including glamping and shepherds' huts, tearoom, gift shop.

- The Gallivant www.thegallivant.co.uk *Lovely coastal hotel near Rye with impressive English wine list*
- Grape Britannia wine shop www.grapebritannia.co.uk *Retail outlet in Cambridge with a huge selection of English and Welsh wines*
- Hawkins Bros www.hawkinsbros.co.uk *Huge list of English wines for sale in person or online*
- The Pig – at Bridge Place www.thepighotel.com *This Pig hotel is near Canterbury in the heart of the Wine Garden of England. British kitchen garden food. I love my Pig visits.*
- Vineyard www.vineyardmagazine.co.uk *The only monthly magazine about viticulture and winemaking in Great Britain. I look forward to my copy each month.*
- Wine Pantry www.winepantry.co.uk *London's original, and exclusively English wine merchant. Founded to promote and support English wines. Wine tours and events.*

INDEX

ACKNOWLEDGEMENTS

Commissioning editor Fiona Holman
Art director Laura Russell
Proofreader Maggie Ramsay
Indexer Angie Hipkin

Pages 2–3 Nyetimber
Page 21 above Rathfinny
Page 21 below Exton Park

Page 66 Cephas Picture Library/Mick Rock
Pages 86–7 Cephas Picture Library/Mick
    Rock
Pages 96–7 Ridgeview
Page 118 Hambledon
Page 137 above Camel Valley
Page 137 below Sharpham
Page 171 White Castle